JOE RIIS

YELLOWSTONE MIGRATIONS

BRAIDED RIVER

FOR MY PARENTS
—JOE

CONTENTS

23	Migration Routes Map
25	Capturing Migrations by Joe Riis
31	The Images of Joe Riis by Gretel Ehrlich
39	PHOTO GALLERY Redefining Home in the American West
55	A New Vision for Yellowstone: An Ecosystem Defined by Migration by Emilene Ostlind
87	PHOTO GALLERY Primordial Paths, Epic Journeys
119	Sustaining Migrations in the Modern West by Arthur Middleton
127	PHOTO GALLERY Barriers & Solutions
147	Epilogue by Thomas Lovejoy
151	PHOTO GALLERY The Photographer's Story: Quiet Persistence
162	Bibliography
164	Contributors
169	Photographer
171	In Appreciation
173	Acknowledgments

Above: Elk congregating in a crucial winter range location, the Spence & Moriarity Wildlife Habitat Management Area, near Dubois, Wyoming **Opposite:** Elk, mule deer, and pronghorn all migrate well outside national park borders and even beyond what is known as the Greater Yellowstone Ecosystem. (Source: *Atlas of Wildlife Migration: Wyoming's Ungulates*, Oregon State University Press ©2018 University of Wyoming and University of Oregon)

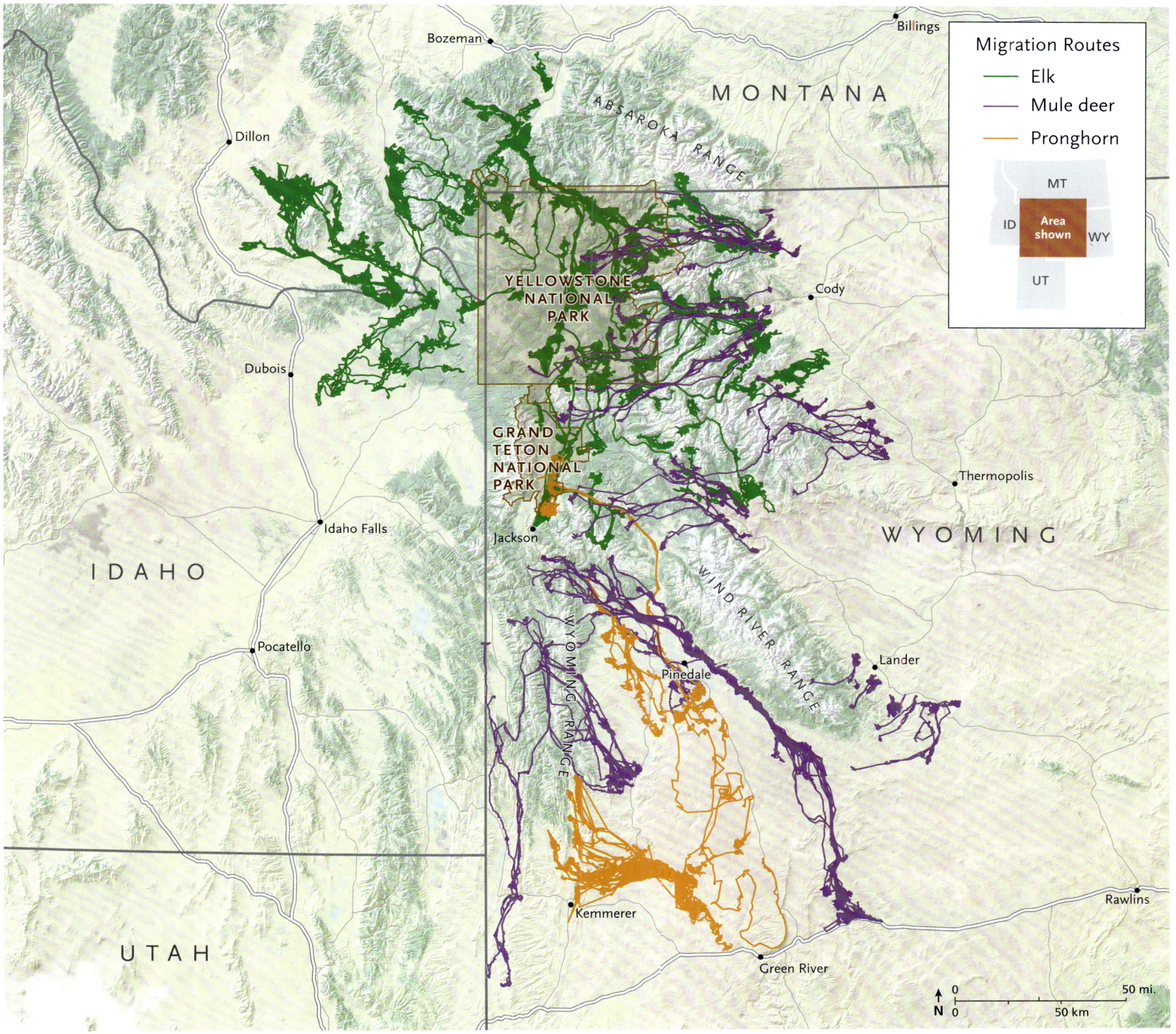

Pronghorn do not typically jump, but they can. Here a small buck pronghorn leaps across a creek swollen from snowmelt.

CAPTURING MIGRATIONS

JOE RIIS

I SPENT THE FIRST eighteen years of my life on the northern Great Plains. My upbringing was out in the wide open, dictated by the four seasons, the ebb and flow of growth and dormancy. I watched the massive movements of birds in the fall and again in the spring. As a kid I always wondered, where did they come from, and where did they go? Birds were my entry into the world of animal movements and migrations, and I later became curious about land-animal migration. But I didn't read of ungulate (hoofed animal) migration until 2003, in Rick Ridgeway's riveting *National Geographic* story (accompanied by Galen Rowell's photographs) about chiru antelope migration in central Asia. The described Chang Tang Plateau was remote and unfathomable to me. I marveled that the summering grounds of these antelope were unknown, the animals' yearly journeys a blank in the human web of knowledge.

Before I was born, my parents lived in Alaska where they developed a passion for wildlife photography. They hauled their equipment with them when they moved to South Dakota. When I was fifteen, I found the box of old cameras in the basement. I practiced with one of them—a Canon F1—in our backyard and became interested in capturing the scene in front of me. Little did I know that I was developing a way to express myself and find meaning.

When I turned seventeen, my father took me to Denali National Park in Alaska to visit his longtime friend and moose researcher Victor Van Ballenberghe. This was my first opportunity on such a trip to focus on wildlife photography. My goal was simply to achieve a sharp and correctly exposed photograph. I saw wolves, moose, bears, Dall sheep, and caribou. After returning home and reviewing my slides, I realized I had great interest in both the photography process and the moose, Dall sheep, and caribou that Victor spoke about. I found them more interesting than the predators that relied on them for food, and I wanted to learn more about their rhythms and way of life.

That curiosity led me to read about the great caribou migrations in the far North and the wildebeest migrations in East Africa. These animals' annual journeys fascinated me, and I started to learn about the idea of wildlife corridors and connectivity. Ecologists like Michael Soulé, Thomas Lovejoy, and George Schaller made this idea mainstream, and photographers like Florian Schulz and Michio Hoshino brought wild animals to life for everyone to see. I was a wildlife biology student at the University of Wyoming and I wanted to get involved in the wildlife corridor movement. Photography seemed like a good way to bring other people in, an avenue for everyone to experience the migrations.

When my friend Emilene Ostlind, who wrote the primary essay in this book, told me about a recently discovered pronghorn antelope migration in western Wyoming, I was very interested. I could not believe there were ungulate migrations like those of the wildebeest and caribou right where I lived. I wanted to see the pronghorn migration for myself, so I searched for photographs. But I found only a few images of antelope behind fences, nothing that captured the essence of migration the way I envisioned it. This was my chance to become a wildlife photojournalist and show a new story.

Like many before me, I dreamed of becoming a *National Geographic* photographer, so I applied for a National Geographic Society Young Explorers Grant to help cover my field expenses—and I got lucky. *National Geographic* and others took a chance and gave me the opportunity to do the unthinkable—photograph the pronghorn migration in Wyoming for the first time. I loaded up my small pickup and drove to western Wyoming, where I would spend the following two years with one mission—figuring out and showing through photographs what pronghorn migration looks like.

Migration is fundamental to many wild animals. It is the process of moving with the seasons to eat and reproduce. Without migration, abundance declines and wildness disappears. In the past hundred years, our human population has gone from less than two billion to more than seven billion people. We dominate the landscape and do whatever we please. The human species is an incredible success story. But other life on the planet suffers the consequences. The American bison has been reduced from thirty million individuals to fewer than twenty thousand free-ranging animals over only a few centuries. The loss of this migration has massively transformed the landscape. However, we have a stronghold of wildness in the Rocky Mountains of the United States—the Greater Yellowstone Ecosystem.

At the heart of this wildness, there is a Greater Yellowstone human community that cares deeply about the wildlife they share the landscape with. The Cody elk herd winters on more than a million acres of private ranchland, managed for both cattle and wildlife. It is absolutely crucial that we view our ranching community as land stewards, because without them, large tracts of open space will become developed and critical wildlife habitat will be lost forever. I have come to care deeply about these people who make a living off the land—their lifestyle, their traditions, and their knowledge of wildlife movements. The outfitter and hunting communities in Cody have also welcomed me, sharing their deep understanding of the migrations. And one family in particular is part of my fondest memories: logger Mark Domek, his wife, Pat Poletti, and their daughters, Callie and Sara, are some of the few year-round residents in the upper Green River Basin. People often refer to them as the Domek herd because they are quick to share stories about their wildlife sightings, promoting the ways of local land stewardship. More often that not, conservation happens because local people see the need to protect their heritage and livelihood. This is why I love the Greater Yellowstone Ecosystem. I feel like I do a small part in helping local people see the beauty and magnificence of the migrations that share this landscape.

In this region, we have a complete assemblage of wildlife centered around Yellowstone, the world's first national park. We still have the abundance and magnificence of animals moving with the seasons. Three of the most incredible and recently discovered overland migrations happen in this ecosystem—the Grand Teton pronghorn, the Red Desert to Hoback mule deer, and the Cody elk.

Through my decade-long journey of photographing these ungulates, I have gained a new understanding and appreciation of the Yellowstone migrations. I believe these animals' annual journeys are what maintain the true wildness of this place.

Capturing the essence of the migrations, however, is challenging. When I began my pronghorn fieldwork, I quickly learned that the animals were reacting to me. My presence on the landscape made it difficult to see and photograph the migration. I realized I had to use new camera technology to document the animals without disturbing or influencing them and their need to move. Motion-activated cameras became my portal to experiencing and discovering what migration looks like in the Greater Yellowstone Ecosystem. Since my first outings, I have spent several years looking for migration pinch points and specific game trails that would be good locations for cameras. Often, I have found hoofprints on trails, evidence of passing, and so the next year I returned earlier in the season to attempt to capture the migration itself. Through this process, I have "seen" the arduous and awe-inspiring journeys that the pronghorn, mule deer, and elk make every year. My hope is that you see these migrations on the pages that follow and are inspired to keep our landscape connected and human communities strong, so that future generations will live in a world as rich as ours is today.

Elk make their way up and down mountains to graze in the high Thorofare Plateau. Located in Wyoming, just outside the southeast corner of Yellowstone National Park, the Thorofare Plateau is one of the wildest and least accessible places in the Lower 48.

CAPTURING MIGRATIONS

While undertaking their strenuous fall migration, elk will sometimes endure deep water, downed trees, blizzards, and ferocious winds.

THE IMAGES OF JOE RIIS

GRETEL EHRLICH

AT FIRST LIGHT, twenty-four pronghorn antelope move to the edge of Wyoming's Green River. They will have already crossed innumerable highways, natural gas fields, and subdivisions, and ducked under barbed wire fences, by the time they join others on their long migration trail. It's May 5, and just barely spring at 7000 feet in western Wyoming, but there is movement everywhere: a pair of swans circles and lands on a half-iced-over pond, mule deer graze alongside the antelope, and a small group of elk trot out of a draw, noses up, and finally settle just above the others. The whole hillside seems to be moving northward, and in the air above, Canada geese, western bluebirds, sage grouse, and a bald eagle circle before they land. Sun comes on strong, and we see the haze of green intensify as snowbanks lose their top crust and meltwater flows from beneath lumpy white mounds. We hear something and turn: behind us a pair of sandhill cranes does its flapping, jumping, bowing mating dance.

I'm here with wildlife photographer Joe Riis, who has photographed all over the world but always comes back to the Greater Yellowstone Ecosystem and Yellowstone National Park, an area that comprises roughly 7 million acres of wild, almost impenetrable land. "That first spring on the path of the pronghorn," Joe tells me, "I realized right away that I wanted to understand the struggle of migration. The intimate details of the life of an animal. Where they walk, where they sleep, when they mate, where they bear their young, what they eat, where they rest, how they avoid predators and human disturbances. I set out four camera traps and moved them often as the animals moved, so that my presence wouldn't deter them. And the images I got back sent me wanting to see more."

Sun is overtaken by a snow squall that turns to rain. Below on the Green River, moose wander through meltwater bogs on large ranches where they will calve and then climb into the high country with their young. In the next few days, cushions of blue and white phlox bloom, and cold-weather grasses such as June grass and alpine fescue appear. We watch the speedy antics of pronghorn running forward, stopping, nibbling at grass, then running back. A sage grouse struts up the hill. Two mallards float by between bits of ice. They stay close to the bank, navigating rocks and tiny rapids, bouncing over what for them are big oncoming waves.

All morning Joe and I tramp through mud and sagebrush, climbing over melting snowbanks, following pronghorn and mule deer. Walking with them day by day, we come to understand that "home" is serial, not a cozy nest, but movement itself, from desert terrain to high alpine forests, peaks, and plateaus and back again, exactly the same way. Between mountain ranges—the Wind Rivers and the Wyoming Range—the sky is swiped navy blue. The animals are looking for forage, a predator-free journey, and a safe place to have their young. By the time they get to Hoback Basin and the Gros Ventre Mountains, the pronghorn will have typically traveled 60 miles, the mule deer even longer—150 miles.

Migration is incremental at first because of the remaining winter snow. We watch the animals walk right up to a 2-foot-deep snowbank, sniff it, and turn around, waiting for it to melt a few feet more. At morning we crouch close to a bend in the wide Green River where pronghorn traditionally cross. A group of about forty or fifty approaches. One female edges forward, sniffs the water, and runs back fast, the rest following her. Ever vigilant—pronghorn are preyed upon by coyotes and wolves—they take their time and finally cross. In some deep places, the water is up to their necks. A fast current pushes them downstream, but they scramble up on the far side, find the trail, and continue on. They have followed this exact path for thousands of years.

To spend a day, a week, a year traveling with Joe Riis in the backcountry is to enter a paradisical green vault—the wild country that spreads out in every direction from Yellowstone National Park would require a lifetime to explore. Every step of the way, we ponder what it will take for humans to exist in this vast territory such that the herds can move through.

As pronghorn follow the snow line, so do the mule deer. We watch five family groups cross the edge of a lake, splash through a river, and climb a wide slope just "coming green." "Their migration was the most surprising to me," Joe tells me. "I guess because deer are a common sight. No one ever bothered to learn about them. But now we know: they winter near the Killpecker Sand Dunes and summer at the very top of the Wyoming Range, as far up as they can go, sometimes grazing alongside a few courageous moose who

While they rarely feed on pronghorn, bald eagles are opportunistic scavengers. This juvenile eagle seized an opportunity to feed on the ears of a pronghorn that died after becoming trapped by downed logs in a fish habitat structure.

also like to go high. [The mule deer] . . . cross more highways and jump more fences than any of the other animals. I have such respect for them. They are smart and careful: they notice my little cameras along the trail—the others don't; and the bucks wait out hunting season and begin fall migration late. Though often caught in snow, they still make it back to the desert."

Looking at Joe's images, I feel as if I'm walking with every animal in every season. Joe says, "I want to let people see what it's like to be an animal, to spend four months of one's life on the move. To encounter weather and obstacles, natural gas fields, downed timber, snowstorms, subdivisions, five-strand fences, rivers in flood. To see this, to get a sense of the mind and life of an animal . . . nothing else is as important to me."

In 2014, Joe joined wildlife ecologist Arthur Middleton to document the migrations of elk herds in the Greater Yellowstone Ecosystem. Guided by Wes Livingston, a Cody outfitter, and his mule, Rosie, they rode 1500 miles with the Cody herd—one of nine distinct elk herds in the large region. "I chose them to photograph because theirs is the most arduous journey," Joe says. They make their way up and down mountains to the high mesas of the Thorofare Plateau, one of the wildest and least accessible areas in the Lower 48, and they persist, undaunted.

Spring migration starts out on two large, historic ranches, the Pitchfork and the Hoodoo west of Meeteetse, Wyoming, in the arid sage-grasslands along the Greybull River. From there, the way leads to a nearly vertical crossing over a 2-foot-wide, knife-edge ridge high on the flanks of 12,100-foot Needle Mountain. Traveling all night, the elk descend to the South Fork of the Shoshone River, reaching it at dawn, and from a wide gravel bar, plunge into rapidly rising waters. Last year Joe watched the crossing: "It was June 8th, and the water came up real fast, carrying huge boulders and whole trees downstream."

When a young elk calved at the edge of the Shoshone River, the herd stopped and waited until the calf could make it across. "That night, when we made camp high above in the trees, the mother and calf edged close to our tents for protection, but a grizzly killed the calf and ate it right there, a few feet from where we were sleeping. . . . She could have taken our cook tent apart, or killed us, but she didn't. She ate the calf and left everything else alone."

Joe noticed that the Cody herd was especially hardy: "They had to be. Their trip was the most arduous. The young calves still have spots in their coats, and the cow elk travel more like mountain goats, capable of dealing with any kind of obstacle: deep water, downed trees, blizzard, extreme heat, ferocious winds. Migration is strenuous. These animals are awe-inspiring."

From Needle Mountain, Joe, Arthur, and their guide, Wes, and his faithful mule, clamber up 4700 feet to another pass. Then the men hack their way with crosscut saws through ten avalanche chutes, some places so steep and congested with beetle-killed downfall that the elk barely make it through. Up the elk go, picking their way along a dusty trail, tongues hanging out, veins bulging. They swim in water over their heads and trudge up another mountain through spring snow. Along the way they care for their young, resting to give calves a chance to nurse. One year in late July, at the top of Fall Creek Pass, Joe saw the entire herd waiting for a cow elk who had broken her leg. He watched her struggle over the pass and

Rocky Mountain elk have the largest antlers of all the subspecies of North American elk.

Riis's images of animals on the move help us to understand, and empathize with, the challenges they face.

continue to limp along. "[The herd] moved slowly that day," he recalls, "waiting for her. They didn't just run ahead. They looked out for each other. They worked together as a herd."

After Fall Creek Pass, the country opens up into grassy plateaus, and at 11,000 to 12,000 feet, the elk finally rest. This is the very heart of wilderness, and they sense it; they thrive. Joe's images of these animals are shockingly fresh and carry a kind of innocence. We feel intimacy with them, almost catching their scent.

While the elk, mule deer, and moose graze on protein-rich high-altitude forbs and grasses in August, grizzly bears eat moths. Having ascended the 11,000-foot-high scree slope of Silvertip Basin, the bears make a small platform to stand or sit on and then dig for moths. "No one knew about this important food source until the mid-1990s," Joe tells me. These Miller's moths are also midmigration, having flown from Nebraska and the Dakotas.

Summer is short at this elevation, over by the end of August when the snow begins. It falls lightly at first but steadily deepens by late September, and the elk, mule deer, and moose have to turn around and descend their vertiginous trail.

The world is its own magic. Each species in the Greater Yellowstone Ecosystem is a wonder to behold: moths and bears, pronghorn, elk, and mule deer, lynx and wolverines, wolves and coyotes, sandhill cranes and swans, plus an extraordinary number of ducks, songbirds, and raptors. The vast territory they traverse is bold and wild, serene and sparkling, a living environment so fecund, so snow- and wind-thrashed, sun-kissed and untrodden by humans, that for a moment, you feel a sense of gaiety and hope that, against all odds, these creatures may survive the rigors of a rapidly heating climate. Joe Riis's images of animals on the move encourage us to develop more tenderness toward life, toward "Otherness." That's when life gets exciting. When we take ourselves off the map and see the liveliness that remains.

To maintain essential wildlife corridors, we have to reimagine housing and fences, rights and privileges. We must work together, even in neighborhood units, to provide room for animals to pass through, to winter on land we aren't using. We can leave gates open in the winter and remove the bottom wire on our fences for pronghorn to squeeze through. We can make our map include not just our houses, towns, and highways but also migration trails, seasonal nesting and denning sites, and pathways across lakes and rivers, meadows and mountains.

In doing so, our strange sympathies, as Ralph Waldo Emerson called them, will have reached deep into the soil and air, into habitat. We will hear and see the animals with whom we must live in a different way. They will bring a sense of joy that we have been longing for but have been unable to define. It's right here. Catch it while you can, cherish it. Joe Riis's photographs are alive with a dynamic co-residency where boundaries are erased, where we learn the meaning of the word B E H O L D.

REDEFINING HOME IN THE AMERICAN WEST

During an extended backcountry trip, Riis, writer David Quammen, outfitter Wes Livingston, and wildlife ecologist Arthur Middleton navigated through patches of deadfall from the Yellowstone fires of 1988—the largest wildfire in the recorded history of Yellowstone National Park.

Opposite: Wes Livingston served as Joe's primary guide through the backcountry, teaching him to ride horses and mules, load pack mules, and find elk trails. His mule Rosie is a peculiar sort who loves campfires. **Above**: Rosie, smiling for the camera.

Above: Once numbering up to thirty million individuals, bison were nearly wiped out by 1889; now public lands support fewer than twenty thousand wild animals, roughly a quarter of those are in Yellowstone National Park.
Opposite: The 145,000-square-foot French Gothic–style monastery of The Monks of the Most Blessed Virgin Mary of Mount Carmel under construction. The monastery is located in the winter range of the Cody elk herd.

While working out of a backcountry base camp tent in the Shoshone National Forest, researching trans-boundary wildlife migrations in the Greater Yellowstone Ecosystem, Joe and Arthur Middleton cooked breakfast and dinner over a campfire every day.

Alden White splits wood for the cook fire at a backcountry camp in the Shoshone National Forest during fieldwork.

Opposite: To fuel his strenuous journey, this grizzly bear will stock up on what sounds like an unlikely fuel source—moths. After ascending a high scree slope like this one, the bear will make a small platform to stand or sit on, then dig for moths. Scientists now know that a grizzly bear can eat up to 40,000 moths in a single day, but the importance of this food source wasn't realized until the mid-1990s. **Above**: Grizzly bear sow and two cubs in snow

Arthur Middleton looks out over a wild and lonesome stretch of the Shoshone National Forest.

Above: Public land is important to hunters, and outfitting and hunting are important aspects of the local economy. **Opposite:** Hunters stay warm by a small fire.

A band of pronghorn migrates through the
Red Hills in the Gros Venture Mountains.

A NEW VISION FOR YELLOWSTONE

AN ECOSYSTEM DEFINED BY MIGRATION

EMILENE OSTLIND

AN EVOLVING YELLOWSTONE

On a spring evening in 2009, I sat on a high embankment in western Wyoming and watched more than one hundred migrating pronghorn antelope. They streamed over a ridge in slanting evening light and trotted downhill toward the cottonwoods and cobbles at the edge of Fish Creek. In early June, as snow melted in the high country, Fish Creek was a frothy torrent of choppy, brownish water tossing tree limbs and foam. One after another the antelope jumped in. Only their heads and necks rose above the water as they were swept along, striking for the far shore. It seemed impossible for their tiny hooves to propel them in that swift water, but several yards downstream each of them reached the far bank and trotted up the gravel bar, dripping and shaking water from its coat before continuing on.

These antelope were midway through a 100-mile-long journey from winter range to summer range. Weighing between 90 and 150 pounds, the same as a small person, antelope are designed to speed across the open sagebrush steppe, not to break trail through drifts or swim rivers swollen with spring runoff. But to migrate, that's exactly what they must do. They had started in the Green River Basin to the south, and over several weeks, they worked their way up the Green River and climbed to a snowy 9000-foot mountain pass in the Gros Ventre Mountains. Their destination was Grand Teton National Park, another day's travel ahead. They would stay in the park through the summer, bearing their fawns and browsing sagebrush to put on fat.

I, too, was midway through a journey, three days into a five-day solo backpacking trip along the remotest section of the antelope migration corridor. In the previous days, I had plunged into thigh-deep snow with a heavy pack weighing me down. I'd watched the thin-legged pronghorn struggle through, a lead doe punching a trail for the rest of the group to follow. I'd seen other animals, too: swans and sandhill cranes, a badger, and even three wolves, retreating from a pronghorn carcass splayed in the snow. I was trying to understand the migration and what compels the antelope to undertake it.

These antelope represent just one group of the many that migrate through the mountainous landscapes surrounding Yellowstone National Park. Deer, elk, moose, bighorn sheep, bison, and antelope all undertake historical, long-distance migrations to survive in a landscape where deep snow closes off the mountains in winter and hot, dry conditions make the lowlands unfavorable in summer. And these hoofed species are just one small slice of the thousands of wild animals, from moths and butterflies to hawks and whales, that migrate around the world. Yet we know little about these migrations, even ones like that of the pronghorn antelope taking place right in the middle of the continental United States.

In 2007, I teamed up with wildlife photojournalist Joe Riis, and we spent several years trying to understand how and why these antelope migrate. We shared what we learned in talks, slideshows, and photo exhibits. We hoped our stories from the migration trail would help more people understand this phenomenon of migration that stitches together a desert gas-field winter range and a protected national park summer range, distant landscapes that seem to have little do with one another. As it turned out, for Joe this pronghorn migration project was just the first step of a much greater journey, one he would follow for many years to longer and wilder and more spectacular migrations throughout the Greater Yellowstone Ecosystem.

From Park to Ecosystem

In 1872, President Ulysses S. Grant signed a bill making Yellowstone the world's first national park. At its inception, Yellowstone National Park was a straight-sided rectangle, roughly 50 miles wide by 60 miles long. Those linear boundaries protected geysers, waterfalls, hot springs, and other geologic features. Inside the rectangle, settlement by people and wanton hunting or fishing were prohibited. The place was meant to "provide for the preservation, from injury or spoliation, of all timber, mineral deposits, natural curiosities, or wonders within said park, and their retention in their natural condition," according to the 1872 act that created the park. This was a relatively creative experiment, to outline a piece of land and set it aside to forever be protected "for the benefit and

Pronghorn that summer in and near Grand Teton National Park undertake one of the longest terrestrial migrations in the Lower 48, leaving the high valleys of the park for Wyoming's Green River Valley, approximately 100 miles south.

Pronghorn migrate south for the winter, often moving before the first snowstorms of the fall.

enjoyment of the people." It established the national park idea and led over time to the creation of hundreds more such protected areas throughout the country and around the world.

Yellowstone's linear borders shifted some over the years, and today its eastern edge better traces the landscape's hydrologic divides. In 1929, Congress added the smaller Grand Teton National Park just south of Yellowstone and in 1950 expanded it to include more forests and parts of Jackson Hole. Over the decades, the lands surrounding Yellowstone and Grand Teton transformed as well. Railways, roads, and eventually the interstate highway system wove through, bringing people, towns, cities, and industry with them. For the most part, that development was held back from park borders by a buffer of rugged, high-elevation forest reserves.

In some areas, settlers pushed out the wildlife, which sought refuge in the forests and two national parks. Meanwhile, as access to the parks improved, the growing number of visitors who ventured into them were increasingly as impressed by the bison, elk, and bears as they were by the geologic features. Today visitors cherish Yellowstone and Grand Teton national parks for the wildlife they harbor, from tiny bats and butterflies to the formidable wolves and grizzlies.

Almost exactly a century after Yellowstone's creation, a scientific idea changed the way that biologists and others thought of the park. In the early 1970s, biologist Frank Craighead applied the term *ecosystem* to the lands surrounding the park. He described a network of national forests and wilderness areas adjacent to Yellowstone that were connected by grizzly bear movements. Over the years, the idea of a "Greater Yellowstone Ecosystem" took hold, though different entities drew its edges in different places. For example, some

traced its boundaries along those of the half dozen national forests adjacent to and surrounding the park. In 1991, geographer Richard Marston and biologist Jay Anderson, focusing on watershed functions and vegetation more than wildlife, described the Greater Yellowstone Ecosystem as those areas within a 7000-foot-elevation contour, a region much larger than what Craighead's grizzlies had defined. Wherever its exact boundaries, today what we call the Greater Yellowstone Ecosystem covers millions of acres of mostly public mountains and forests, spanning about 300 miles from central Wyoming into eastern Idaho and southwest Montana.

Thanks to the ecosystem idea, we no longer accept that the interiors of Yellowstone and Grand Teton national parks are safe regardless of what happens outside their boundaries. The several federal land-management agencies responsible for this area—the National Park Service, US Forest Service, US Fish and Wildlife Service, and Bureau of Land Management—coordinate their activities and policies to better take care of the system. Thinking of this larger region as a place that animals move through for survival constitutes a more complex understanding of Yellowstone than its early protectors could have realized when they drew that first rectangle on a map.

The Migration Link

Nearly half a century after adopting the Greater Yellowstone Ecosystem concept, it is time again to transform our thinking about this place and consider the conservation implications. And to grasp this new shift, we have to look to migration. Somewhere in the middle of Yellowstone National Park's web of life are the ungulates, hoofed mammals built to travel long distances season to season.

Six wild, native ungulate species—pronghorn, mule deer, elk, bison, moose, and bighorn sheep—all migrate through the Greater Yellowstone Ecosystem to take advantage of lush high-elevation forage in the summer and to escape from snow at lower elevations in the winter. These animals are built the way they are, with stout, muscular bodies and durable hooves, so they can migrate. Their migrations take them tens of, and in some cases well over a hundred, miles between summer and winter ranges.

In the spring, as the days lengthen and the snow begins to pull back from the valleys, foothills, and high mountain meadows, the ungulates follow its receding edge, climbing in elevation. They nose up against the drifts that melt under the spring sun to make way for new green shoots. The young, bright, fast-growing vegetation is the most nutritious of the year.

For some animals, the spring migration lasts as long as eight to ten weeks. This is not just a quick trip from one place to another, but a critical part of the year when they put on fat after a lean winter and produce milk for newborn calves, fawns, or lambs. Some species, such as elk, give birth to their young along the way, and the offspring make the journey with the adults. Other species, like pronghorn, wait until arriving at the summer range to give birth. Either way, keeping pace with the spring green-up enables them to find the best forage and gain the most weight. In fact, scientists have clearly shown that the animals who stay behind, opting not to migrate (known as "residents"), do not gain as much weight as their migratory counterparts.

The animals' summer retreats are deep inside the Greater Yellowstone Ecosystem's parks and forests. In these cool, wet, high-elevation ranges, the migrants find rich green grass and browse well into the summer. The nutritional content of that forage translates to fat, the fat translates to fast-growing offspring and energy to survive the winter, this survival translates to reproductive success again the following year, and that translates to robust populations.

As the days begin to shorten in the autumn and the first snow falls in the high country, the animals then retrace their journeys back to the lowlands. The winter storms push them, like a huge exhalation, out toward the edges of the Greater Yellowstone Ecosystem and beyond, back to the lower elevations and drier landscapes where they can escape deep snow and eat the grass and shrubs that grew through the summer. Now that improved GPS collar technology is helping biologists map these migrations for the first time, we are starting to understand how the park animals rely on distant deserts, valleys, and agricultural lands for half the year. Many winter ranges are below the 7000-foot contour, beyond the accepted edges of the Greater Yellowstone Ecosystem.

Wildlife populations share these winter ranges with human communities, including towns like Cody, Pinedale, and Jackson, Wyoming, as well as more rural areas such as natural gas fields in the Green River Basin and huge working ranches in the Bighorn Basin. Many of these lands are still wild and wide open, providing safe winter habitat for the wild animals that venture there. But these areas are also vulnerable. Year by year, development encroaches, fragmenting the landscape and in some cases, cutting off migration corridors. New gas rigs and pipelines eat up the sagebrush. Private ranchlands sell to developers, who chop them into subdivisions. Rural sprawl cuts off the very passages that wildlife desperately needs. Roads and highways, towns and cities, reservoirs and fences interfere with the migration corridors and winter ranges on the flanks of the Greater Yellowstone Ecosystem. In some places, such development compromises the wintering ungulate populations, and that loss echoes deep inside Grand Teton and Yellowstone national parks when fewer animals return to the summer ranges. Even as the system appears strong and protected, it is crumbling at the edges.

Yellowstone National Park's creators in 1872 could not have fully realized how these hoofed migrators link the park interior to the outside, or the barriers these animals would one day have to navigate to make their journeys. As we learn how these migratory herds need more than the forests and parks in the Greater Yellowstone Ecosystem, we can't conserve migration routes in the same way that we initially conserved national parks. We can't just draw another rectangle and set it aside for wildlife. We need a new model, a new way of interacting with the landscape, a way to start knitting our own existence more closely with the other species that also rely on these lands for survival.

PRONGHORN: PASSAGEWAY IN PERIL

Outfitters and ranchers working at the edges of the Greater Yellowstone Ecosystem have grasped the magnitude and significance of the ungulate migrations for decades, but they see only bits and pieces of the journeys. They don't know exactly where the animals that appear in one high-elevation valley each spring have come from or where those passing by ranch gates will spend the winter. Biologists have long tried to tease out that mystery. Before the advent of GPS collar tracking technology, linking individual animals on a summer range with winter habitat took some creative approaches.

In 1985, researchers trapped more than three hundred pronghorn antelope in southwest Wyoming and fitted them with colored neck bands printed with a letter and number. Over the following four years, they added more animals to the study group and monitored their movements, recording observations of them in different places throughout the seasons. Most of the antelope migrated around the Green River Basin, but a few showed up in Grand Teton National Park, more than 150 miles north. This was some of the first sound evidence of the great distances that migratory antelope cover.

In July 1993, Hall Sawyer, who had recently finished his master's degree at the University of Wyoming studying elk in the Bighorn Mountains, along with Wyoming Cooperative Fish and Wildlife Research Unit leader Fred Lindzey and Wyoming Game and Fish biologist Doug McWhirter, captured thirty-five female pronghorn in and around Grand Teton National Park and fitted them with VHF radio collars. Such radio collars emit a frequency on a unique radio channel that carries for up to a few miles. Over the following two years, the researchers used special antennae to locate the collars from fixed-wing airplanes about once a week during the spring and fall migrations. In the spring of 1999, the researchers also followed pronghorn tracks on the ground through a portion of the migration corridor. That let them draw a rough map of the route, a jagged series of line segments linking Grand Teton National Park to the Green River Basin in the south.

In 2003 another team of researchers, including Grand Teton National Park biologist Steve Cain and the Wildlife Conservation Society's Joel and Kim Berger, again captured pronghorn in the

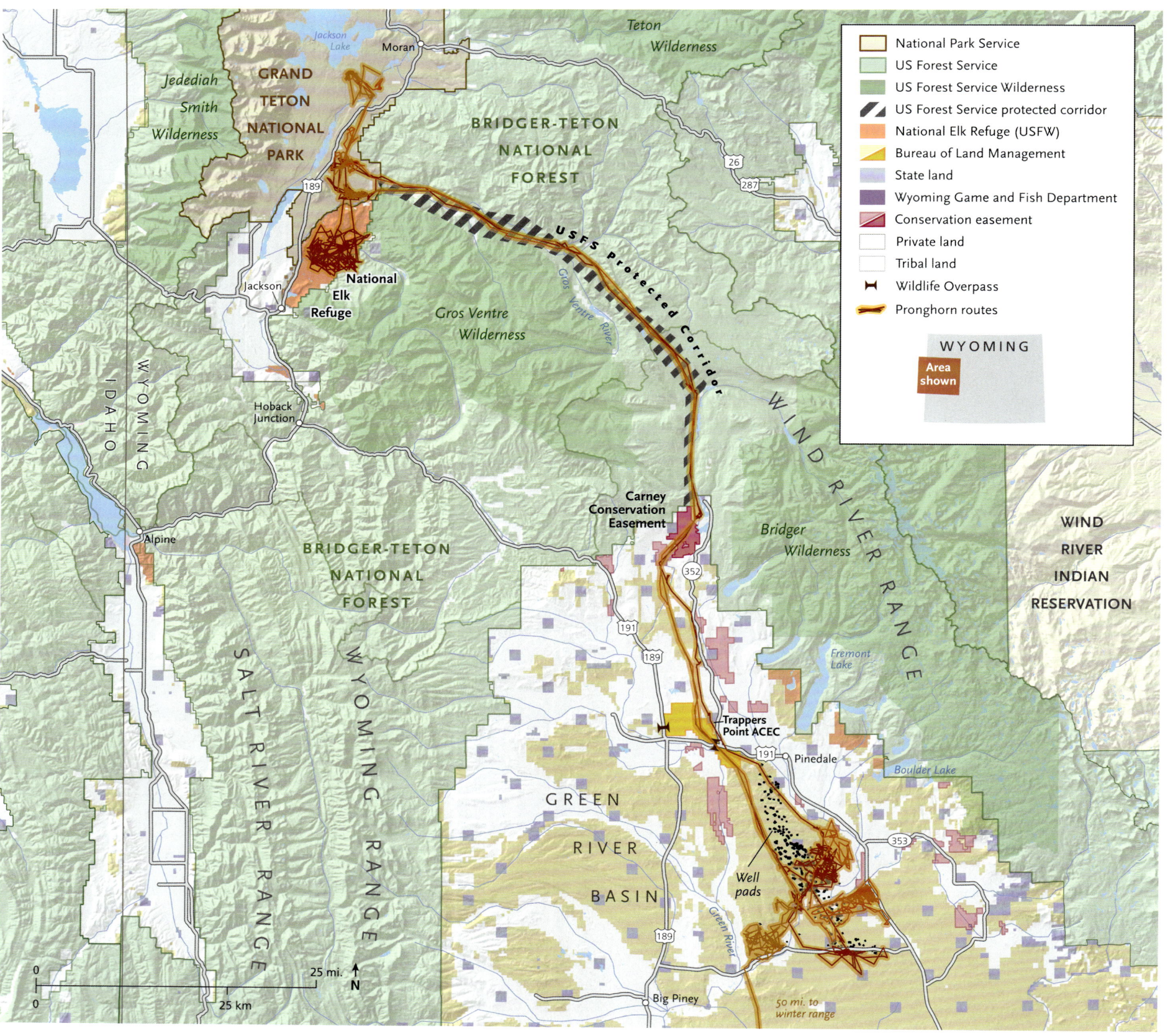

Designated in 1891, the Shoshone National Forest was the first national forest in the United States. Stretching from the Montana state line south to Lander, Wyoming, the Shoshone holds portions of the Absaroka, Wind River, and Beartooth mountains. Yellowstone National Park sits on its western border.

park, this time fitting them with GPS collars. GPS collars communicate with satellites to fix locations on a given time interval and store the waypoints on a computer chip. After two migration seasons, the collars automatically dropped off the animals' necks and the researchers collected them. The data showed all the animals leaving the park in the fall, moving through an identical sequence of drainages and narrow trails over the mountains, and spreading out across the winter range before returning to the park along the same paths the following spring.

Cumulatively, these studies provided a window into the secrets of ungulate migration that no one had seen before. They showed how each fall, three hundred or so pronghorn antelope set off on a long, dangerous trek, crossing a narrow band of cliffs and a high mountain pass, fences and ranches, subdivisions and energy fields, to winter in the southern desertlike basin. There is no other path for these animals to leave the park. One female pronghorn migrated a full 170 miles to the south, stopping only when Interstate 80 blocked her path.

The researchers called this the longest land-mammal migration in the continental United States, second longest in the Western Hemisphere after that of Arctic caribou. And it was imperiled. In one area the animals snuck through a cluster of cabins between steep forested slopes and a river. Should that housing development expand, it could cut off the migration, the researchers warned. In another place, the animals darted across a busy highway in one quarter-mile-wide area, dodging trucks and minivans, and several pronghorn were killed each migration season. A web of fences spread across the antelope migration corridor and a web of natural-gas roads and drill pads spread across their winter range.

Migration in Danger

All over the world, various wildlife species undertake long seasonal journeys that create a fabric of life, stitching landscapes together. And all over the world, these migrations are blinking out as new development eats up habitat, fragments once-intact regions, and blocks corridors. As migrations disappear, the species that rely on them decline, further degrading the surrounding natural systems.

The starkest example of human-built infrastructure interfering with the western Wyoming pronghorn antelope migration was the Trappers Point highway crossing (see "A Solution for Both People

The path of the pronghorn is long and dangerous, requiring the animals to negotiate both natural and manmade obstacles as they complete the longest land-mammal migration in the continental United States. (Source: *Atlas of Wildlife Migration: Wyoming's Ungulates*, Oregon State University Press ©2018 University of Wyoming and University of Oregon)

LIFE—AND DEATH—IN A MIGRATION CORRIDOR

When Joe Riis set out to document the western Wyoming pronghorn migration, he quickly realized he needed help from the people who knew the country and its wildlife best. His most important allies were people working and living in the migration corridor. In particular, these included Pat Poletti, the since-retired postmistress at the Cora Post Office, and her lumberjack husband Mark Domek, who runs a sawmill. They live in a little log house Mark built overlooking the pronghorn migration corridor where it traces the Green River, one of the few families that stay year-round this high up the river valley. Their daughters, Sara and Callie, about Joe's age, are true mountain women who grew up backpacking the Wind River Mountains, running a dogsled up the Green River valley in winter and making plaster casts of wolf and bear tracks from the muddy trails just beyond their back door.

The Domeks keep binoculars on their kitchen table to watch the wildlife flowing along the Green River—moose and sandhill cranes and elk and trumpeter swans and, of course, antelope. Mark would call Joe in the spring as soon as he saw the first migrating antelope, and Joe would drive out from South Dakota to stay in the Domeks' sheep wagon that they keep as a guesthouse.

In January 2009, Joe was at the Domeks' when a large group of antelope that had stayed late in Grand Teton National Park started their migration. Deep snow already buried the upper Green River valley. Sara called from the Moore Ranch up in the mountains to say she had seen the late migrators, so Joe and Callie cross-country skied up the valley to find the antelope. They counted seventy-four, as much as a quarter of the Teton herd, slowly working their way south through the deep snow.

The next morning, Joe heard a commotion in the kitchen and found Mark setting up a dusty old spotting scope he had pulled

Sara, Pat, and Callie Domek prepare lunch at their "sheep camp" near their home in the Upper Green River basin near the pronghorn corridor.

out of a closet. He trained it on the meadows a mile away across the Green River. There were the late migrators, plodding single file through the snow, and they were not alone. As Joe, Mark, Pat, and Callie watched, three huge wolves loped down from the forest, almost floating above the snow on their snowshoe-like paws. They began picking off the antelope, one, two, three, dragging them down. The antelope surged, lunging and struggling, until the front of the line reached a packed trail and raced out of sight to the south. Meanwhile, ravens and eagles descended on the carcasses, and coyotes, which Joe said looked like little rats compared to the wolves, appeared out of nowhere.

Over the following days, Wyoming Game and Fish Department staff packed a snowmobile trail for the antelope and monitored their progress. Several animals froze to death or died from exhaustion, and the rest stumbled down to their winter range another 30 miles south. Meanwhile, the wolf kills were etched into Joe's mind, emphasizing the importance of the migration not only for the antelope themselves but also for the whole cast of other wildlife species that depend on them for sustenance.

and Wildlife"). So named for the mountain men who rendezvoused here in the 1830s, Trappers Point is an undramatic, yellowish hill about 6 miles west of Pinedale, Wyoming. It's a narrow strip of land shaped like the neck of an hourglass, outlined by rivers on either side. The migrating pronghorn, like sand falling through the hourglass, all cross the highway in one quarter-mile-wide area, at the narrowest spot. In the 1990s, archaeologists working here unearthed seven-thousand-year-old bones of adult and fetal antelope, suggesting that the spring migration has passed by this very spot for thousands of years.

This was a dangerous place for the migrating antelope. Drivers surged over the crest of a blind hill at 65 miles per hour just where the migration corridor intersected the highway. Animals crossing in the evening or at night were especially hard to see. The traffic included not just rural ranchers and hunters but also workers and tanker trucks from the massive natural gas fields to the south.

Furthermore, fences lined both sides of the highway. Antelope rarely jump fences, and when one occasionally does, it risks twisting a foot between the two top wires, a death sentence. Fences that reach down to the ground, especially woven wire sheep fences, can block antelope all together, trapping them far from the winter range as snow piles up. Joe's camera trap on the right-of-way fence captured video of animals pacing back and forth, searching for a gap to crawl under. In one shot, a buck tangled a horn in the wire, lunging and shaking his head until it popped free.

The state transportation department had identified Trappers Point as a dangerous place for drivers, too, and put up an elaborate system of signs that were supposed to flash when animals ran onto the highway and triggered them. But the signs frequently misfired, and even with the extra warnings, vehicles still collided with wildlife, both antelope and mule deer. It was not uncommon to find a pronghorn carcass crumpled on the highway shoulder.

These dangers did not go unnoticed. After the collar studies illuminated the main threats to the pronghorn migration, including the highway crossing at Trappers Point, conservation organizations and land-management agencies started to take action. Grand Teton National Park managers realized that protecting the wildlife that came to the park each spring depended on cooperation from landowners and land managers far beyond the park boundaries.

In 2008, the Bridger-Teton National Forest amended its forest plan to prevent future development in the northernmost 45 miles of the antelope corridor. The Jackson Hole Conservation Alliance organized volunteer days to remove unnecessary fences from the migration path. The Green River Valley Land Trust rebuilt hundreds of miles of fence on ranches along the Green River to make it easier for antelope to crawl underneath. In 2009, the Bureau of Land Management designated 9500 acres surrounding Trappers Point an Area of Critical Environmental Concern, putting it off-limits to natural gas drilling and other surface disturbance. In 2010, the Conservation Fund placed a 2500-acre conservation easement on a ranch that overlapped the corridor along the Green River to prevent any future subdivision or development there.

But despite the monumental conservation efforts, the antelope corridor remained vulnerable to existing and future fences, highways, and other obstacles. The gas boom brought people and money to this part of Wyoming, and once-open agricultural land was being cut up into housing developments. The mountain views and abundant wildlife attracted more people who built vacation homes and hunting retreats. Fences, houses, driveways, corrals, traffic, and powerlines slowly and steadily cluttered the corridor. Already a subdivision had blocked one half of the hourglass neck at Trappers Point, where Joe had watched pet dogs harass a group of antelope trapped between the right-of-way fences. Just as biologists were starting to understand these migrations and conservation groups were taking action, the threats were intensifying. There was more to be done. Because this pronghorn migration is so long, so precise, and crosses so many landownership boundaries, protections in one area might be fruitless if the corridor became blocked somewhere else.

MULE DEER: AN UNDISCOVERED MIGRATION

Following his 1998 VHF study of the Grand Teton National Park pronghorn migration, Hall Sawyer continued to study deer and antelope movements all around the West. He earned his PhD developing an analysis method to identify the most critical resting places within corridors, and he became a respected authority on wildlife migration. For one research project, the Bureau of Land Management (BLM) office in Rock Springs, Wyoming, hired him to delineate the home range of a population of mule deer that wintered just north of Interstate 80. Everyone believed these were resident deer that, if they migrated at all, likely stayed within an area spanning about 40 miles of the Red Desert from the interstate up to the toe of the Wind River Mountains in central Wyoming.

In January 2011, Sawyer collared forty mule deer in their winter range. The collars both emitted a VHF radio signal and recorded a satellite location every three hours. (If Sawyer had known where those deer were going to take the collars, he would have invested in the more expensive ones that transmit data in real time.) He could periodically locate the VHF signals from the collars using an antenna, but he would have to wait two years for the collars to drop off the animals to download the GPS location data.

Later that spring, Sawyer hired a pilot to fly over the Red Desert and search for the radio signals. Sawyer was sitting in his office in Laramie, Wyoming, when he got a text: the deer were missing. The pilot had scanned the area from the capture site to the Wind Rivers and found only a couple of the deer. Where could the animals be? Sawyer suggested the pilot keep searching for signals as he made his way back to the airport in Pinedale, 100 miles to the north.

Along the way, the pilot started detecting signals from more collars. The deer had migrated up the western flank of the Wind Rivers, traveling much farther than anyone had expected.

Over the following two years, Sawyer located the deer a few more times, and their story came into focus. In western Wyoming, close to the center of the continental United States, where wildlife has been studied endlessly, where avid hunters stalk deer year after year, where people have observed and swapped stories about them for generations, was a migration no one knew about: a 300-mile round-trip mule deer journey, even longer than the antelope migration Joe had photographed.

When Sawyer collected the collars and downloaded the GPS data, he found that most of the deer traveled in the spring to a mountain valley known as the Hoback Basin and then retraced their route in the fall back to the Red Desert. This wasn't a fluke—the animals repeated the same route the following spring and fall.

Sawyer realized that, like the antelope, the deer face all kinds of obstacles on their journey. But unlike the antelope, these deer don't enjoy the security of a national park at their summer range or a protected corridor through national forest land at one end of their migration. While the deer's summer range lies high in a national forest at the edge of the Greater Yellowstone Ecosystem, their fall migration takes them through mostly private and BLM lands, places vulnerable to development, to a winter range far beyond the Greater Yellowstone Ecosystem's accepted boundary. Sawyer thought that if he could identify the threats along this migration corridor, he could give conservationists, land managers, ranchers, and others a chance to protect some of its most vulnerable spots and keep it intact.

After he delivered his report to the BLM, he teamed up with the Wyoming Migration Initiative, a fledgling effort to improve understanding and conservation of big-game migrations started by Wyoming Cooperative Fish and Wildlife Research Unit leader Matt Kauffman. Sawyer wanted to share the mule deer migration he had discovered with people beyond the few scientists who might read about it in a journal article. He envisioned a multimedia approach that would tell the migration story to a wide audience via photos, video, maps, and text. He hoped such an effort would help people understand how and why the deer moved through this diverse landscape. That's where Joe came in.

Pinch Point

On an early November weekend in 2013, Sawyer and I drove from Laramie to meet Joe at the cabin north of Cora, Wyoming, where he was living while he photographed the fall mule deer migration. Joe and Sawyer had flown the length of the corridor in a helicopter the previous spring, counting fences and touching down on buttes

Dr. Hall Sawyer, a biologist with Western Ecosystems Technologies, Inc., inspects a GPS collar, learning that the doe mule deer wearing it had traveled more than 150 miles one-way, to the Hoback Basin, before the collar dropped off.

to scan the landscape. They surveyed in an afternoon what it took Joe and me three migration seasons on foot to learn about the antelope migration. At the cabin, Joe showed us recent photos on his computer, and we talked over a plan for the following day. We kept the woodstove stoked through the night and woke up to 6 inches of fresh snow in the morning.

The mule deer migration is longer and more complicated than the antelope migration, and it raises larger questions about conserving migration in the Greater Yellowstone Ecosystem. Mule deer, which make their homes in forests and deserts, are an iconic species throughout the Rocky Mountain region. They are bigger than pronghorn, weighing up to 250 pounds, and more elusive. Their defining characteristics include a dark cap on their foreheads, large forking antlers coveted by hunters, and enormous ears, nearly as big as their heads. Whereas antelope have a curiosity that, if anything, draws them toward a camera, mule deer tend to startle at the click of a shutter. They bound away, all four hooves leaving the ground at once, a movement that lets them navigate rugged terrain. Hunters respect them because they are difficult quarry.

While mule deer are abundant across western North America, they are in decline range-wide. Wyoming reported a high of around 578,000 mule deer in 1991; over the following quarter century, that number dropped by more than a third to less than 375,000. Similar decreases have taken place in Colorado, Utah, Idaho, and other western states. Game managers are still trying to understand why and curb the decline. Migration is part of the answer.

A yearling buck mule deer kicks up some dirt while migrating south for the winter in the wide-open sage country of western Wyoming.

Like pronghorn, mule deer have honed their survival strategies to fit western landscapes. Their numbers depend on their ability to migrate between sheltered, low-elevation winter ranges and lush, high-elevation summer ranges so that they can gain fat and bear healthy young. While mule deer jump fences more easily than antelope, they are sensitive to highways, subdivisions, and energy development that can cut off their access to the best forage, especially in winter and spring when they need it the most.

On that November weekend, Sawyer took us to see a key 400-yard-wide pinch point where new development could easily block this corridor. Each spring and fall, four thousand to five thousand deer sneak through a gap between the outskirts of Pinedale, Wyoming, and 8-mile-long Fremont Lake, whose upper reaches are buried deep in a glacial canyon in the mountains. We looked across a stretch of shallow, flat water at the lake outlet. Thousands of hoof prints, pointing south, pocked the smooth, muddy bottom.

For now, the deer were finding their way through, passing a campground, boat docks, parking areas, subdivisions, and an 8-foot-tall fence meant to stop elk from reaching the lower agricultural lands in the winter and transmitting disease to livestock. The exact parcel of land spanning the lake outlet was free from development, but as private land, it could be sold, subdivided, and built up with houses at any time. A new fence or a few lakeside homes with driveways and dogs would close off this piece of the migration forever.

We returned to the truck and over the course of the day worked our way south, stopping to check Joe's cameras and see other creek and road crossings. The farther we got from town, the wilder and more remote the country, the fewer the fences and roads. We didn't see any deer, but their trails were clear—well-trodden paths striking out into the sagebrush, aiming for the distant horizon.

When Science and Photography Work Together

Sawyer and Joe unveiled *The Red Desert to Hoback Mule Deer Migration Assessment* with a presentation and reception hosted by the Wyoming Migration Initiative in Laramie in April 2014. The fifty-page publication included not just Sawyer's description of the corridor and Joe's photographs but also detailed maps of the whole route; the maps showed the complex landownership (dozens of private landowners plus the State of Wyoming, Wyoming Game and Fish Department, BLM, US Forest Service, and US Fish and Wildlife Service) as well as management designations on the public lands, including active oil and gas leases. The final chapter highlighted the top ten areas of concern for conservation along the Red Desert to Hoback mule deer migration. Number one was the bottleneck at the Fremont Lake outlet.

Joe's photos of deer walking in long lines through snow, wading creeks, and swimming the lake outlet brought the data and maps to life. Joe needed the scientists to show him the skeleton of the migration, and the scientists needed Joe's photos to put flesh on the bones. When they worked together, both the science and the photography became stronger.

A four-minute video with Joe's footage of deer splashing through the lake outlet and Sawyer describing the mule deer migration went viral, tallying millions of views in the first months after it was posted. When high-level administrators at the US Department of Agriculture saw the film, they invited Sawyer and Matt Kauffman from the Wyoming Migration Initiative to travel to Washington,

Mule deer prefer a habitat that is a mix of sagebrush, aspen, and forest.

DC, and brief them on western ungulate migrations. Joe, Sawyer, and Kauffman also gave more talks closer to home. Joe's photos hung in galleries and coffee shops, and Sawyer distributed copies of the migration assessment around the state. Increasingly, migration corridors were on the minds of citizens, hunters, ranchers, land managers, policy makers, and others throughout the Greater Yellowstone Ecosystem and beyond.

And piece by piece, even as new development threats arose in the migration corridors, new conservation successes were happening too. In 2011, the Wyoming Department of Transportation designated $9.7 million to build wildlife crossing structures at Trappers Point and along a 12-mile stretch of highway to the west. John Eddins, an engineer and mule deer hunter from Rock Springs, led the charge.

Eddins had orchestrated construction of underpasses for deer on another Wyoming highway with resounding success. But while mule deer readily used the box-shaped tunnels, antelope, which rely on their eyesight and speed to stay out of harm's way, were reluctant to enter the passageways. Eddins proposed building an overpass for antelope at Trappers Point. He showed that savings from averted car wrecks and wildlife deaths would more than justify the cost of construction in just over a decade. And he cited the overwhelming response to Joe's photos of the pronghorn migration as a clear sign of public support for the project.

This was something no one had tried for the species before. Would the antelope know how to use the overpass? As is often the case, construction was late, and in October 2012 when the first migrating pronghorn arrived at Trappers Point, contractors were still working.

Still, the antelope knew exactly what to do. A motion-triggered video camera showed 120 antelope (and a single mule deer) hardly breaking stride as they ran onto the bridge, heading south. The

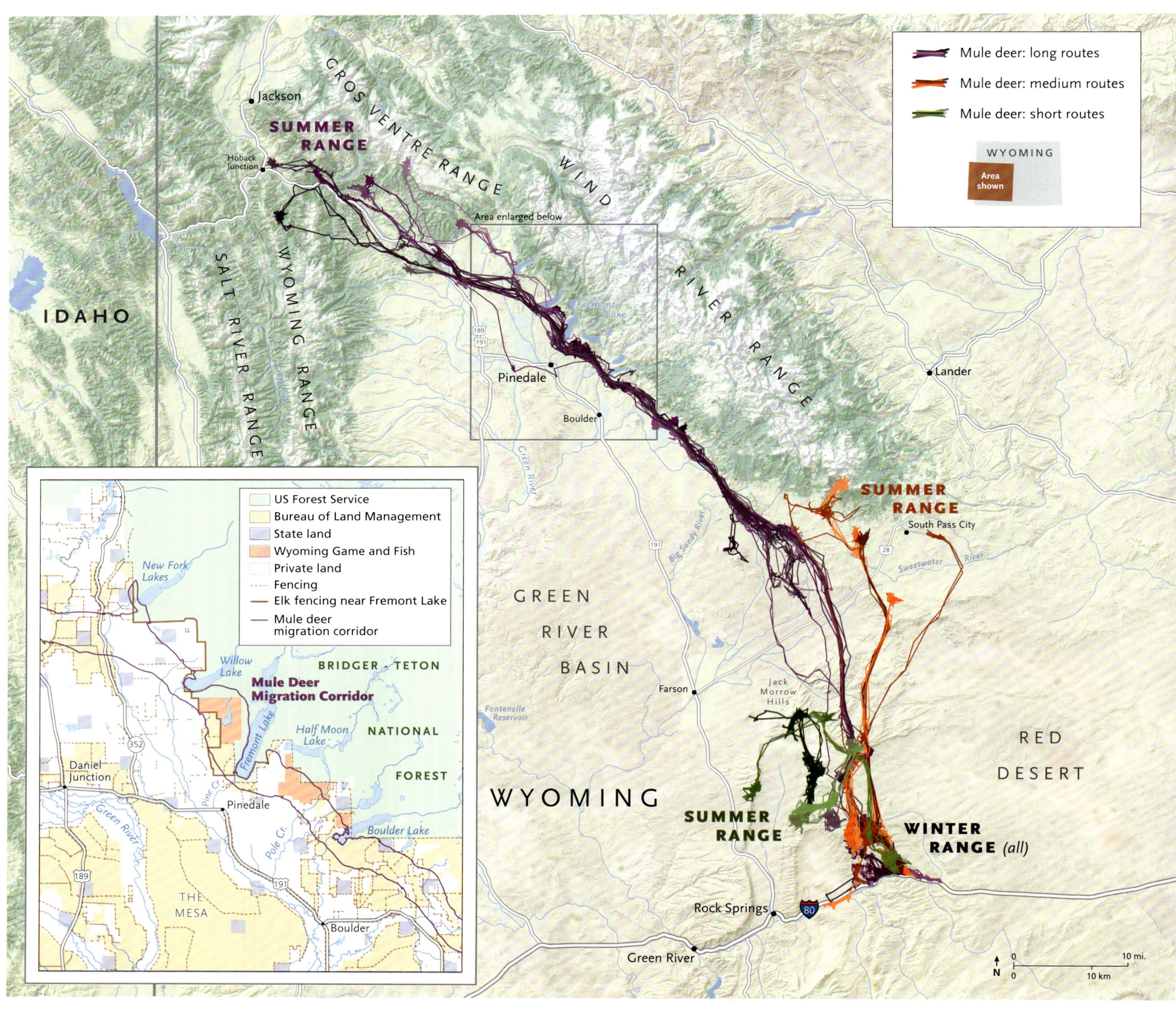

contrast to earlier scenes of stressed animals pacing the fence and running from trucks and dogs was stark. The most dangerous place in the path of the pronghorn was suddenly much safer and easier to cross than it had been for generations.

The transportation department contracted with Sawyer to monitor use of the overpass, and in the first three years following construction, he found that mule deer crossed over nearly 60,000 times and pronghorn crossed over 20,000 times. These animals included not just the three hundred antelope that summer in Grand Teton National Park but thousands of others on their way to and from scattered summer ranges in the north. Also notable, vehicle collisions with pronghorn at Trappers Point dropped to zero.

The spring of 2015 brought another conservation victory for migration in the Greater Yellowstone Ecosystem. The very parcel of private land spanning the Fremont Lake outlet had gone up for sale and a dense housing development was planned. Armed with the mule deer migration assessment, and working swiftly and quietly behind the scenes, Luke Lynch, Wyoming director for the Conservation Fund, made an offer and launched a campaign to raise the $1.7 million listing price, plus another $400,000 to remove migration obstacles. He closed the deal in April 2015, just weeks before losing his life in an avalanche while backcountry skiing in the Tetons. After his death, the Conservation Fund carried out his vision. They moved the elk fence out of the migration path before donating the land to the Wyoming Game and Fish Department as the Luke Lynch Wildlife Habitat Management Area, ensuring that it will stay protected for wildlife into the future. The number one conservation concern for the Red Desert to Hoback mule deer migration was ticked off the list.

Despite these significant conservation actions, both the Teton pronghorn migration and the Red Desert to Hoback mule deer migration remain vulnerable. In fact, nowhere in the United States is a long-distance wildlife migration corridor fully protected. Some biologists believe that highways and other infrastructure have already cut off dozens of historical ungulate migration corridors in the Greater Yellowstone Ecosystem. No one had mapped them, so no one knew where the animals were moving to or from, and the migrations just disappeared without notice.

And development continually presents new threats, especially in a place like the Greater Yellowstone Ecosystem, where huge swaths of protected public land make for scenic views and robust wildlife populations, which attract people to the patches of available private land. New subdivisions, fences, reservoirs, and highways can still sever migrations without anyone realizing what is happening. Unplanned housing sprawl is a surefire way to block a migration corridor. Conservation in this ecosystem will require new ideas, a new model for humans to share a landscape with migratory wildlife.

Sawyer's work, and that of other biologists in the Wyoming Migration Initiative and with other organizations, is essential to this model. GPS collar maps of wildlife migrations provide hard evidence to inform planning and conservation efforts, making it easier to plan passageways for wildlife and harder to let migrations vanish unnoticed. Due in part to the Greater Yellowstone Ecosystem's relatively small human population and vast network of public lands,

Mule deer migrations range from short to long distances; the 150-mile-each-way Red Desert to Hoback migration (shown here in purple) surprised researchers. The mule deer on this long migration face many challenges, including a pinch point between the town of Pinedale and Fremont Lake. (Source: *Atlas of Wildlife Migration: Wyoming's Ungulates*, Oregon State University Press ©2018 University of Wyoming and University of Oregon)

These glacial potholes on private ranch land (Carney Conservation Easement) in the Upper Green River Basin lie within the pronghorn and mule deer migration corridors.

many of the world's best long-distance land-animal migrations still take place in this area. There is still time to study them, to tell their stories, and to keep them intact for the future.

ELK: THE ULTIMATE CONNECTORS

In 2012, another migration researcher sought Joe's camera-trap expertise to help share a poorly understood migration story unfolding in the Greater Yellowstone Ecosystem. Arthur Middleton had earned his PhD at the University of Wyoming studying elk under the direction of Matt Kauffman. As Middleton prepared to start a postdoctoral fellowship at Yale, he dreamed of creating a huge data set and map that would describe all nine of the major known elk migrations that link Yellowstone National Park to surrounding lands. Of the nine hundred thousand or so Rocky Mountain elk in North America, some twenty thousand migrate through the Greater Yellowstone Ecosystem. He recruited Joe, and they put together a proposal to compile existing GPS data from some elk migrations, collar additional elk, set camera traps, and importantly, follow an elk trail in person to its remote summer range deep inside Yellowstone National Park's backcountry. In addition to the data set, the pair proposed to create a film and exhibit.

The duo beat out twenty other applications from around the world to win the inaugural $100,000 Camp Monaco Prize, created to promote study of and communication about Yellowstone's biodiversity. His Serene Highness Prince Albert II of Monaco bestowed the award on them in Cody, Wyoming, in September 2013, one hundred years after the prince's great grandfather had stayed at a hunting camp in Yellowstone National Park. Looking ahead to the next century, Joe and Middleton would use the elk migration story to redefine how the world thinks about Yellowstone.

In part, their collaboration echoed that of Ferdinand V. Hayden, the geographer who led the 1871 expedition to Yellowstone to gather data and map the region so he could build a case for protecting it as a national park, and William Henry Jackson, the photographer who joined the trip to document the area's splendors for all the world to see. As a scientist, Hayden realized that his work would never reach beyond his academic world to shift public consciousness unless he could provide images and a greater story. And like Hayden, Middleton wanted not just a catalog of data but the experience to go along with it. Middleton knew that following the elk along their migration trails himself would make him a better scientist and help him understand the purpose and complexity of the migrations.

To build his map, Middleton needed location data and much more: climate and drought information, cow-to-calf ratios in each herd, wolf and bear kill locations, body condition of the collared animals. He started compiling information on seven of the nine migrations from researchers in Wyoming, Idaho, and Montana, along with his own data from his PhD work on the Clarks Fork herd. Sparse data existed for the Cody and Wiggins Fork herds, so he bought GPS collars and hired helicopters and net gunners to capture elk in those herds. Pretty quickly, the Monaco Prize money was spent, and he and Joe launched into fund-raising to cover the rest of the project expenses, which would eventually add up to $1.5 million.

Middleton's purpose for the whole project reached beyond the migrations, too. Just as biologists in the 1970s had changed the thinking about how Yellowstone National Park worked for conservation when they linked it to the Greater Yellowstone Ecosystem, Middleton and Joe's work, almost a half century later would make another transformation.

From his earlier PhD work, Middleton knew that in winter, the most sensitive time of the year, many of the Yellowstone elk populations rely on private lands beyond the Greater Yellowstone Ecosystem, below the 7000-foot contour and the national forest boundaries. And he knew that emotions about wildlife run high. Some ranchers fear and hate wolves and grizzlies. Some have had their herds quarantined by the US Department of Agriculture after elk passed disease to their livestock. Some struggle to make their ranches break even, much less earn a profit from year to year. And worse yet, those ranches that don't make ends meet sometimes sell off swaths of land for development.

Middleton saw an opportunity in the huge intact ranches that harbor wintering elk in the Cody area. He knew that even those ranchers who hate predators tend to love the country and its wildness. He knew that communities need the Greater Yellowstone Ecosystem for economic, spiritual, and cultural reasons. The ecosystem needs the elk to graze its meadows and feed its predators and scavengers. The elk need to migrate in and out of the ecosystem to be fat and productive and thriving. And in a fitting circle, the migrations need the human communities to keep corridors open and share the winter ranges. Middleton hoped to transform thinking about conservation in this wild and working landscape to protect both the elk winter ranges and the people who take care of those lands.

All these elements would build toward Middleton's greater vision for the project. He wanted elk migration to carry the people surrounding the Greater Yellowstone Ecosystem into a new era, one less about property lines and vitriol over predators and more about aligning to protect a landscape that so many people care about in different ways.

"I want for the elk migration story and the broader migration story to be somehow restorative," Middleton told me. "That would be my dream."

By restorative, he meant repairing relationships through the whole system, among the ranchers and biologists and conservationists and wildlife managers and outfitters and hunters and others, as well as between the people and the land. For that to happen, Middleton needed both scientific data and pictures. Joe's photos would give the migration story its power.

Journey of the Cody Elk

"Elk are the ultimate connector across the landscape," Middleton told me. "We can't think of people and nature and ranches and wilderness as these completely separate things. They're right next to each other. These animals depend on all of them all at the same time."

To get a sense of elk migration, one must, like Joe and Middleton did, follow one migration from beginning to end. The pair narrowed in on the Cody herd. In winter, some four thousand to six thousand Cody elk graze hayfield stubble, find shelter in creek bottoms, and enjoy relative protection from wolves and bears on the Pitchfork Ranch, Hoodoo Ranch, Brown

Thomas Meadow Ranch, Valley Ranch, J Bar 9 Ranch, TE Ranch, Ishawooa Mesa Ranch, 99 Ranch, and Deseret Ranch which range from 500 to 255,000 acres each. As Middleton described it to me, the Cody herd "seemed archetypal in the sense that it's moving between these huge, storied ranches and the deepest wilderness and remotest place in the Lower 48."

Come spring, the elk leave the ranches, headed toward the lush meadows of Yellowstone National Park, some 40 miles away and a few thousand feet higher in elevation. This migration is shorter than those of the antelope and deer, but much more rugged. The elk are made for it. Up to four times the size of a mule deer, their 400- to 1000-pound bodies are pure muscle. They plod steadily along, weaving their way up and down steep mountainsides, and cover tens of miles in a single day.

From the ranches, the elk head up the open Greybull River valley toward Boulder Pass, skirting below 12,000-foot Carter Mountain's rock spires. Now they are inside the Washakie Wilderness Area. They work up a high valley called Boulder Basin and, incredibly, ascend through snowfields almost to the 12,100-foot summit of Needle Mountain, a formidable buttress of gray stone faces. There, thousands of years of passing elk hooves have worn grooves into the volcanic rock.

From this aerie, the elk drop more than a vertical mile down to the South Fork of the Shoshone River, which cuts a deep canyon between Needle Mountain and the Absaroka Range. The lush canyon floor is the opposite of Needle's bare alpine precipices. In springtime, the river churns at banks thick with head-high willows and chokecherries. Cottonwoods and lodgepole pines crowd the edge of the river, and the water pulls them in and hurls them into massive log jams down the canyon.

To get here, Joe and Middleton horse-packed 7 miles up the river. They brought an inflatable kayak to ferry their gear across the water, too high with spring snowmelt for the horses to cross, and they set up a camp. They sent the horse packer away so the horses wouldn't startle the elk or interfere with the migration, and they stayed at this camp for weeks at a time each spring during their two years of tracking the Cody herd. Joe made a photo blind on the riverbank. And waited. And waited. For days, nothing happened.

They weren't the only ones. A grizzly and a couple of black bears knew the elk were coming, too. At the blind, the rush of the water was too loud to talk over, and one morning Joe's sheet-white face and gestures let Middleton know something was behind him. Middleton turned to see a grizzly wake up in the hemlocks just 10 yards away. It ambled toward the water, more interested in elk than humans.

Finally, one morning as Joe watched from the blind, the first elk—cows with young calves—arrived. They walked onto the river cobbles, looking across the white-tipped waves to the far shore. And then, one by one, they plunged into the water and swam, the current sweeping them downstream until they pulled themselves, dripping, up the opposite bank.

As the migration continued over the following weeks, bears snagged unfortunate calves. The carcasses fed not only the black and grizzly bears and their cubs but ravens, eagles, foxes, and other scavengers. Joe described a dozen elk rib cages strewn along the trail by the end of the migration. "That adds a whole layer of bear awareness," he told me. He carried both bear spray and

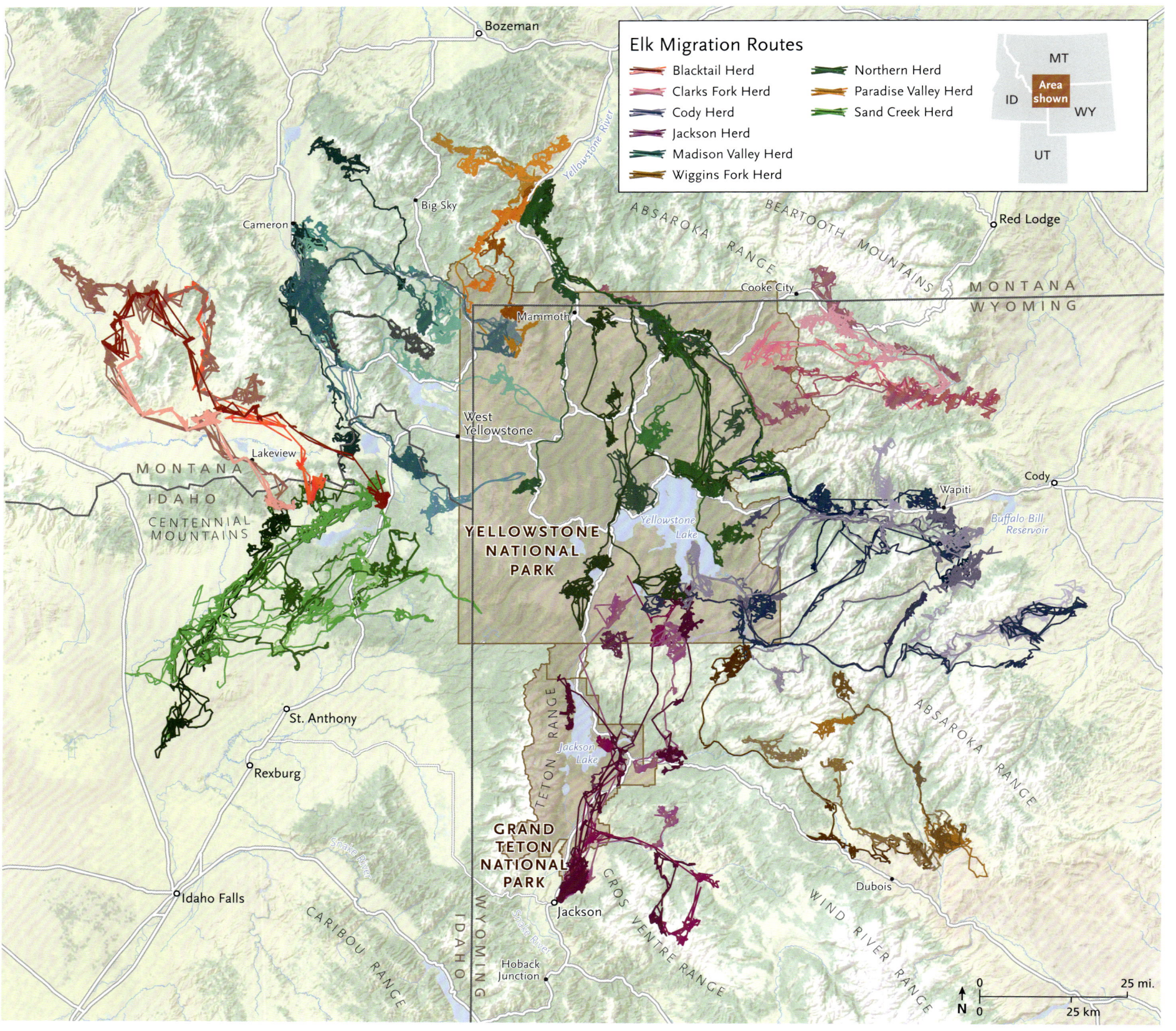

a large handgun at all times, keeping them within reach even while he slept.

Beyond the river crossing, Joe later followed the elk up Fall Creek, a steep-sided drainage choked with brush, trees, and fallen trunks at the edge of the Teton Wilderness Area. The 5-mile climb to 11,000-foot Fall Creek Pass required two day's work, clearing downed trees for the horses and mules.

The pass at the top gives a panoramic view of the Greater Yellowstone Ecosystem. From this high point, long, grassy ridges point to the Thorofare Plateau and the high elk summer range inside Yellowstone National Park. Joe could see the Bridger and Crazy mountain ranges in Montana, the Bighorn Mountains 100 miles east, as well as the much closer peaks of the Absaroka Range and even the Tetons to the southwest. He imagined the web of migration corridors fanning out around him, sustaining the herds that relied on this place to raise their young and linking Yellowstone to the world beyond. As Joe looked out from the pass, he felt this place was somehow . . . restorative.

Yellowstone's Heartbeat

By the end of 2015, Middleton's elk map was taking shape. He had collected about four million GPS locations from 340 elk collars, including those he had put on Cody and Wiggins Fork elk. The map shows the nine elk migrations radiating out from Yellowstone National Park through national forests into Idaho, Montana, and Wyoming. Some of the elk travel well beyond the accepted boundaries of the Greater Yellowstone Ecosystem, into surrounding private lands. In fact, they span a region five times the size of the park.

One can't look at this map without thinking of veins and arteries. Yellowstone National Park's interior is the beating heart and the elk are the blood, surging in each spring, pulled by the suction of grassy meadows where they can grow fat and produce rich milk for their calves; surging back out in the fall, in all directions, pushed by snowstorms and lengthening nights to the sheltered, less snowy lowlands. Before this map existed, the migrations were invisible. The map provides an X-ray view into the machinations of the Greater Yellowstone Ecosystem. It shows an essential driving element of the ecosystem. The elk map depicts how the landscape needs migration.

If elk are the lifeblood, pumping in and out of the park, the ecosystem is the body that that lifeblood sustains. The elk feed the bears and wolves. Their carcasses feed the eagles and foxes, which hunt the fish and meadow voles, which eat the insects and seeds. The bears and wolves and eagles and foxes and elk inspire and mesmerize the ranchers and school kids and scientists and hunters and park staff and park visitors, teaching them about vivid, living systems of biodiverse wildlife. The hunters and visitors pay for gas and hotels and tour guides and outfitters. The gas stations and hotels and restaurants sustain the local communities. The communities are full of people like you and me who wake up in the morning and find themselves in a world rich and vibrant and invigorating because of those elk, knowing that without the elk migrations it would be poorer.

Nine different herds of elk migrate within and beyond the Greater Yellowstone Ecosystem. (Source: *Atlas of Wildlife Migration: Wyoming's Ungulates*, Oregon State University Press ©2018 University of Wyoming and University of Oregon)

Leaning out of a Robinson R44 helicopter and firing a net is the first step in fitting this cow elk with a GPS collar—a process that takes only minutes.

Wes Livingston, a wildlife capture specialist, releases a cow elk fitted with a GPS collar. The orange straps in his hands were used to tie the legs together while he worked the animal.

"Science isn't the limiting factor for conservation in the greater Yellowstone and a lot of places," Middleton told me. "What is the limiting factor? It's basically people's will and ability to care more about the natural wonder in their backyard than about the differences between them. What acts on that is not science. It's the story."

A Vision for the Greater Yellowstone Ecosystem

Joe had found that story on the trail of the antelope, and on the trail of the mule deer, and he was searching for it on the elk trail. He sought an image that would show not just elk, not just the landscape, but the very essence of the migration. He thought the photo might be a procession of huge bulls in snow, leaving the park for their ranchland winter ranges, and he set cameras with such an image in mind. Or maybe it would be one of the bears feeding on an elk carcass, showing how elk nourish the species around them.

In August 2015, Joe and an outfitter were descending Fall Creek from the elk summer range on the Thorofare Plateau in Yellowstone's southeast corner. Joe led a string of four mules, and he could hear their hooves striking rocks, the jangle of bridles. A cool breeze stirred the high-altitude summer air. This was a month-long trip, and on the way out he was collecting the camera traps he had set on the way in. One camera was on his mind, the last one, still a few miles ahead. It was screwed to a stump near the bottom of the Fall Creek drainage.

In Joe's mind, that camera was positioned to illustrate all the elements of the elk migration. It was beside a steep section of the migration trail, aimed through a 10-foot-wide gap in the conifers toward a distant mountain face. The summer before a grizzly had swiped the camera to the ground just hours after Joe set it up. It had lain facedown in the dirt while the entire elk migration walked past. This summer he hoped for better luck.

Joe's years of work pursuing three great long-distance migrations through the Greater Yellowstone Ecosystem built up to this moment. Over time, he had learned the migration trails just as the fawns and calves learn them. By following the antelope and mule deer and elk, he had started to understand these migrations as journeys to sustain the species, journeys motivated by the very folds and textures and shape of the land, by the grasses and wildflowers and shrubs, and by the winter storms and the summer rain.

In the late afternoon, Joe and the outfitter reached the camera. It was intact. Joe took the camera out of its plastic box. Standing on the mountainside, he shaded the screen on the back with his hand to scan through the images.

That's when he saw the heartbeat of the Greater Yellowstone—not bulls in snow or predators, but the procession of cows with tiny calves at their sides, climbing from the South Fork of the Shoshone toward the high pass, their mouths open, dust hanging in the air.

This, Joe saw, was the photograph to match Middleton's map, an image to convey what migration means for the landscape and everything connected to it: the small calves, bawling and crying out to their mothers on their first journey into the Thorofare Plateau summer range, following the ancient trail from the lowlands into Yellowstone National Park. Those little calves, just a couple of weeks old, scared and exhausted but so full of life, give hope. If they can make this journey, then surely we can take care of the landscapes they need to do it. And if we can take care of the calves and their landscapes, there is hope that we can restore these places for ourselves as well.

Of the roughly 900,000 Rocky Mountain elk in North America, some 20,000 migrate through the Greater Yellowstone Ecosystem.

CAMERA TRAP TRIALS

Before Joe Riis embarked on his mission to photograph the western Wyoming pronghorn migration, he searched for images of migrating ungulates. The few he found either showed the animals from the backside, running away, or were taken with long lenses, zoomed in from a great distance. To make intimate, close-up images and avoid frightening the animals or interfering with the migration, Joe decided to set motion-sensor-triggered camera traps. A camouflage waterproof box protects a high-quality digital camera, secured to a tripod or other mount. When a migrating antelope or other animal comes close, it triggers a motion sensor, tripping the camera shutter and essentially taking its own photo.

Joe uses wide-angle lenses. For these photos to work, an animal has to be within a couple of feet of the camera. Joe carefully frames the shots, placing the cameras close to narrow trails and arranging them to capture the nearest animal, the animals trailing behind that one, and the surrounding landscape. He strives to convey what he calls a "migration moment." With pronghorn, Joe thought a river crossing would be his best chance to tell the migration story in a single shot.

He discovered—initially by their tracks and then by waiting and watching—where the antelope crossed the Green River. To capture a photo there, he needed several elements to align: the

framing of the image, animals in the water, the light, the camera settings, the timing. He had to predict how far the current would push the antelope and where they would climb out of the river, as well as how much to expose the frame given changing daylight.

In the spring of 2009, he set a camera just a few inches above the water in a tangle of driftwood at a bend in the Green River. A few days later, when he returned to change the batteries and check the memory card, he found a sequence of ghostly, blurred photos. Antelope had triggered the motion sensor at sundown, which resulted in long-exposure photos in the low light. The animals appeared as transparent paint strokes across each image.

Joe adjusted the camera settings, repositioned the shot, and waited a few more days. This time when he returned, he found photos of antelope swimming the river under gray, cloudy skies. Water drops on the lens blurred the images. He adjusted the camera again, put in fresh batteries, and left it.

This pattern continued for weeks, producing thousands of images that weren't right. The bulk of the migration passed. Joe was running out of time. Finally, one evening he checked the camera and saw what he had been wishing for.

Antelope in the river, swimming toward the camera, their necks and ears lifted forward as if to pull them through the current. In the foreground, one doe, close to the lens, steps out of the water, droplets pouring off her neck and legs. Bright, low evening sunlight glows on her tawny coat and glints in her eye. Her ears are perfectly silhouetted against the deep blue sky. Her shadow, a profile of her face, is cast onto a boulder sticking out of the water beside her. The image portrays a string of animals with a singular purpose, on a necessary mission that has been woven into their very being for eons. They are antelope because they migrate, and they can't be antelope without their migration.

PRIMORDIAL PATHS, EPIC JOURNEYS

Pronghorn bounding through a snowmelt-swollen creek
in the Bridger-Teton National Forest

Opposite: Pronghorns are herbivores, subsisting on grasses, flowering plants, sagebrush, and other prairie plants. The amount of winter forage available is the factor that often limits their survival. **Above**: Weighing up to 250 pounds, the mule deer's defining characteristics are a dark cap on their foreheads, enormous ears, and, on bucks, large, forking antlers coveted by hunters.

A POPULAR PASS

Six ungulate species—pronghorn, mule deer, elk, bison, moose, and bighorn sheep—all migrate through the Greater Yellowstone Ecosystem to take advantage of lush high-elevation forage in the summer, and to escape from snow at lower elevations in the winter. These animals are built the way they are, with stout, muscular bodies and durable hooves, so they can migrate.

Eagle Pass lies on the south boundary of Yellowstone National Park. Nearby Eagle Peak (11,372 feet) is the highest point in Yellowstone National Park. Over a period of two days, as fall turned to winter, three ungulate species, as well as a bear, moved through this particular pinch point.

Doe mule deer (September 28, 11:00 a.m.)

Bull moose (September 29, 10:40 a.m.)

Bull elk (October 1, 5:50 a.m.)

Young black bear (October 1, 11:30 a.m.)

A band of pronghorn crosses the Green River near Pinedale, Wyoming. Adult females (does) typically lead the way when the group crosses rivers or other dangerous places.

Opposite: The bulk of the Cody elk herd's migration takes place outside of Yellowstone National Park. The terrain they cover is five times the size of the park. Some elk travel beyond the Greater Yellowstone Ecosystem and into surrounding private lands. **Below:** Elk cross the south fork of the Shoshone River during their spring migration.

As with pronghorns, when mule deer cross streams along their migration route, female deer often lead the way and bucks follow.

Grizzly bear moving through a pass in the Absaroka Range

The Hoback Basin, nestled between the Wyoming and Gros Ventre mountain ranges, serves as the summer range for mule deer, marking one end of their lengthy migration to and from the Red Desert.

Weighing between 90 and 150 pounds, pronghorn are designed to speed across the open sagebrush steppe.

A band of pronghorn migrate south as the first winter storm of the season settles in. Pronghorn prefer open areas where they are able to see far and run fast; much of their migration corridor is this sort of wide-open sage-steppe habitat.

Along their migration route, pronghorn move as swiftly as they can through thick patches of willows, like this one, where their visibility is limited.

Bighorn sheep can cover tens of miles in a single day.

A bighorn sheep climbs a high pass in the South Fork of the Shoshone River drainage, near the southeast corner of Yellowstone National Park.

Swimming through fast-moving rivers is one of the most dangerous natural challenges that pronghorn face.

Above: Elk navigate through the high mountain passes of the Absaroka Range in the Shoshone National Forest during their fall migration.
Opposite: Sunlight catches Thorofare Peak in the Absaroka Range.

Each spring, thousands of elk from separate herds migrate from their individual winter ranges in the Greater Yellowstone Ecosystem to high elevation ranges nearer to the heart of Yellowstone National Park. These migrations depend on access from the ecosystem's farthest foothills to its deepest, mountain wilderness. Here they travel in the high country of the Shoshone National Forest.

Yellowstone's migrations help to keep the Greater Yellowstone Ecosystem's natural diversity intact. After the reintroduction of wolves, this region regained its full historical complement of vertebrate wildlife species. Every large wild mammal that inhabited it when Europeans first arrived in North America can be found in the Greater Yellowstone Ecosystem today. This aerial view shows Hidden Creek in the Thorofare region of the Shoshone National Forest.

Above: As snows begin to fall each autumn, some three hundred to five hundred animals gather and begin their trek back to their Upper Green River valley winter range. They must reach and cross the hydrographic divide between the Green and Gros Ventre rivers before snow blocks their way. **Opposite:** The pronghorn's path through National Forest and National Park land is unobstructed, but once they reach a housing development at the Forest Service boundary, pronghorn must begin to navigate human development. While they can make their way through the current development, it is uncertain how much more they can handle.

Petroglyphs at Legend Rock

SUSTAINING MIGRATIONS IN THE MODERN WEST

ARTHUR MIDDLETON

ABOUT 75 MILES SOUTHEAST of Yellowstone National Park, by a creek near the edge of what we now call the Greater Yellowstone Ecosystem, there stretches a long wall of Native American petroglyphs. Many of them depict hoofed animals, or ungulates, like elk, deer, and pronghorn antelope. They are simple tan figures, made by scraping away the darker skin of the rock. Some date back ten thousand years.

Of all the stories and creatures they knew, why did generations of Native people highlight the ungulates in particular? For me, after a decade of wildlife research in the Greater Yellowstone Ecosystem, the site—now called Legend Rock—seems like a tribute. The Shoshone and other tribes depended on ungulates for food, tools, and shelter. I imagine they saw these animals as the lifeblood of the landscape.

Today's science is helping us see just why ungulates are so important to the Greater Yellowstone Ecosystem. Six species—elk, deer, pronghorn, moose, bighorn sheep, and bison—migrate seasonally between low valleys and high mountains. Migration helps them survive harsh winters and regain fat in summer. My own first inklings of the migrations' importance came from my study of the Clarks Fork elk herd, which winters north of Cody, Wyoming, and summers in the upper Lamar River valley of Yellowstone National

Arthur Middleton leads a horse through the backcountry while following the Cody elk herd's migration route from start to finish.

Park. The migratory elk that followed the spring wave of fresh green grass into the mountains could gain up to 15 pounds more fat than their nonmigratory counterparts. That extra fat supports breeding and survival. In short, it supports abundance.

Since then, my colleagues and I have been learning that these nourishing migrations are more common among the ungulates around this large Yellowstone region than we previously recognized.

Recent tracking studies show that many of the herds travel annually anywhere from 25 to 150 miles. One especially striking discovery came in 2012 when my colleague Hall Sawyer, a wildlife biologist who has long worked on these issues, documented a 150-mile mule deer migration from southern to northern Wyoming —which we now know as the longest land migration in the Lower 48. More recently, I worked with agency biologists around the region to aggregate GPS collar data and to create the first detailed map of the vast network of elk migrations in and out of Yellowstone National Park. That we can still make such advances in a well-studied ecosystem suggests there is much more to discover—and that migrations are even more essential to this region than we knew.

The implications of these discoveries for our understanding of this ecosystem's natural and human economies are enormous. Altogether, tens of thousands of animals in dozens of herds generate an abundance that sustains the park's famous wolves, bears, and scavengers and attracts millions of tourists and many thousands of hunters annually. In turn, these visitors pay local businesses, like hotels and restaurants, for their services. In towns around Yellowstone—places like Bozeman, Montana, and Cody and Jackson, Wyoming—large statues of elk, moose, and sheep adorn institutions such as banks. In the autumn, many locals greet each other with, "Get your elk?"

Once, trying to explain the migrations' importance to a friend from New York City, I found myself posing a simple thought experiment. Imagine if the national park boundary were a wall, blocking the movements of elk, deer, pronghorn, and other ungulates. Deprived of access to lush summer grass or to shelter from winter snow, many herds would die out. Predators and scavengers would go hungry. Outfitters and guides would go out of business, and hotels and restaurants would struggle. The ecosystem would slowly unravel.

There will never be any such wall, but the endgame will be much the same if we keep clogging and severing migration routes around Yellowstone. We know from many studies that poorly planned housing subdivisions, oil and gas developments, and fences and roads pose the greatest threats. Hall Sawyer's research has shown that over fifteen years the development of a high-density natural gas field in the southern Greater Yellowstone Ecosystem cut in half the number of deer in one migratory herd. Other studies have shown how deer move more quickly through developed areas, skipping important stopover areas where they used to rest and feed. In many cases, the major impact of development is not just the conversion of open space to infrastructure but animals' avoidance of the remaining habitat because of new noise and truck traffic.

Fences and roads illustrate the complexity of the threats to migratory ungulates. Crossing a single fence or highway may not seem to matter much, but many herds must cross dozens of fences and roads, twice each year. The energy costs and survival risks can add up, threatening what my colleague Matt Kauffman calls the "incremental loss of migration hypothesis." In his thinking, the benefits of migration are slowly outweighed by cumulative human-imposed costs—a new fence here, a new road there, another subdivision over there. Populations decline and, ultimately, the animals even lose their memory of traditional routes. What is most alarming is that the slow pace of declines and the fuzzy multiplicity of their causes may render the loss of herds or migration routes nearly imperceptible. Will we even know what we're losing?

That is why Joe Riis's migration photographs are so important. They are an extraordinary visual chronicle of our discoveries—a Legend Rock for our time. By pinpointing the migration trails and shadowing the herds, Riis has made a unique contribution: he has brought us along as the animals traverse mountains and plains, allowing us to see moments otherwise impossible to witness. We see elk crossing roaring rivers that are swollen by spring snowmelt; pronghorn squeezing between subdivisions; mule deer crossing highways and human-made lakes. Riis's images reveal migration as the very essence of these animals. But at the same time that he elevates these animals' beauty in our eyes, he shows us, viscerally, the hurdles we put in their way. His photos insist that we ask, "How can we make these animals' lives easier? How can we help sustain these migrations in our modern world?"

Riis's work has helped illuminate the critical importance of an informed and engaged public. An all-too-common illusion is that scientists (like me) generate new knowledge and then pass it over to government decision makers, who apply it. But progress in wildlife conservation and management depends heavily on excited, informed people *asking* decision makers for change. It was such public interest that in 2008 prompted the Bridger-Teton National Forest to protect the 40-mile Path of the Pronghorn from future development. It was public interest that in 2011 led the Wyoming Department of Transportation to build highway overpasses and underpasses to give deer and pronghorn safe passage. And it was public interest that in 2015 nudged the Wyoming Game and Fish Department to designate migration corridors as critical habitat in the context of future development plans. The images in this book have played a role in many of these changes.

The Wyoming Migration Initiative's (WMI) Red Desert to Hoback migration assessment, which evaluated the ecology and conservation of the newly discovered 150-mile mule deer migration set a new standard in this regard. More recently we've seen the special power of science and photographs working together. In leading the report, Hall Sawyer and the WMI went far beyond the conventional scientific process to couple the excitement of discovering an extraordinary migration with a rigorous assessment of the threats to its persistence, presenting the whole with beautiful graphic storytelling. Soon, conservationists, hunter groups, and state and federal agencies came together to tackle the number-one threat to the migration: a potential 300-acre housing development that could have fully blocked the animals' journey. Science together with public outreach, not one or the other alone, inspired this action—and will inspire others in the future.

Still, significant challenges lie ahead. Case studies from now-famous pronghorn and mule deer migrations have stimulated innovations in science, management, and policy, but these migrations are but single threads in a fabric that blankets the Greater Yellowstone Ecosystem and much of the American West. How are we going to conserve multispecies networks of migrations at the scale of a whole ecosystem? And if the Greater Yellowstone is a microcosm of the West, how can we scale up and speed progress across nearly half of North America?

Clearly, part of the answer is to plan our future economic development to minimize impacts on wildlife connectivity. As we discover and map new migrations, we can use migration assessments to guide protection of key habitats. Where development is inevitable, we can avoid the most important habitats—like core winter and summer ranges, stopover areas, and high-use migration corridors—and favor technologies that minimize noise, traffic, and other disturbance. We can use fencing retrofits and install overpasses and underpasses to reduce risks and stressors along major corridors. Pursuing these activities on a large scale will require that local, state, federal, and tribal agencies, in the Yellowstone region and beyond, integrate

Improved GPS collar technology has shown that many winter ranges for Yellowstone's wildlife are found in the deserts, valleys, and agricultural lands outside the mountain ranges and national forest boundaries, and even beyond the accepted edges of the Greater Yellowstone Ecosystem. This female elk's collar will allow researchers to study her migration path.

migration more fully into policies and management than they do at present.

The greatest challenge of all, however, is not merely technical. It is social and cultural. When people first see the nine elk migrations of the Greater Yellowstone Ecosystem on a single map, set against the patchwork quilt of landownership, they are often overwhelmed by the complexity. Most elk in the course of their year travel through lands delineated as parks, wilderness areas, multiple-use forests, state trust lands, private ranches, and the Wind River Indian Reservation. Though the animals themselves pay little attention to who is in charge on each plot, the humans certainly do. A herd that summers out of sight and mind in Yellowstone National Park may winter on neighboring ranches, eating valuable grass, drawing predators to mix with cattle, and carrying infectious diseases. The migrations are not just geographical transects of the landscape, then; they are also transects of a highly complex social landscape fraught with deep tensions and conflicts. Many of the interested parties—local, state, federal, and tribal governments; ranchers; environmental groups; hunters; oil companies—are skeptical of one another. Indeed, some among them became sworn enemies during the era of reintroducing large carnivores such as wolves and grizzlies to the ecosystem.

This is why the real challenge in sustaining wildlife migrations and conserving habitat connectivity is our ability and willingness to work together across boundaries that are not only physical but also cultural and ideological. In my earliest years in the Greater Yellowstone Ecosystem—a time that some call "the wolf wars"—no challenge seemed more daunting than this one. But as I have traveled the trails of the elk in years since, I have met diverse people who are moved by what we have been learning—even by the simplest of our maps. I have met ranch owners and managers who have told me that our learning broadens their view of their responsibilities on the ranch. I have met outfitters who have told me that it's time to instill in the hunters they guide a new ethic—that these animals are much more than just trophies—and that the migration routes should be protected for the benefit of their industry and livelihoods. To me, these and many other recent conversations suggest that a cultivated love of place, and of the natural wonders in Yellowstone's backyard, may be strong enough to bridge the old divisions and transcend ideological boundaries.

And it's in this way that I have come to feel that the migrations of the Greater Yellowstone Ecosystem have as much to offer us as we can offer them. For decades, conservation has progressed by playing on the differences among us instead of looking to our many shared values. In the future that I see, conservation will doggedly uncover, strengthen, and renew connections among people that are required to sustain connections in nature. That is the lesson that all the wildlife depicted so beautifully in this volume teaches us.

At Legend Rock, not all the figures are ungulates. Some are hybrid creatures—antlered humans—perhaps symbolizing the acute dependence of Native inhabitants on these animals. Centuries and millennia later, are we all that different? The elk, the deer, the pronghorn give us enjoyment, inspiration, sustenance, and livelihoods. Our fates may be linked as tightly as they ever were.

Arthur Middleton and Wes Livingston and their dogs Jack and Dee.

BARRIERS & SOLUTIONS

Why did the mule deer cross the road? Because its ancient migration routes dictated this path long before any road—inclucing this state highway in western Wyoming—existed. Now the mule deer's migration routes require them to navigate many manmade hazards across a combination of private property and public lands.

Opposite: Natural-gas drilling rigs encroach upon a pronghorn migration route near Pinedale, Wyoming.
Above: Two pronghorn bucks near a drilling rig

Pronghorn antelope gathered near Trappers Point along US Highway 191. This photo was taken in 2008, prior to the construction of the highway overpass at the same location. Identifying and preserving migration routes for conservation of these animals is critical as development, growth, and fragmentation of habitat dramatically increase across the region.

Above: Two pronghorn cross a highway outside Pinedale, Wyoming before the overpass at Trappers Point was constructed. **Opposite:** A mule deer buck approaches a 7-foot-tall fence designed to keep elk away from private lands.

A SOLUTION FOR BOTH PEOPLE AND WILDLIFE

Before this wildlife bridge was constructed, the Trappers Point highway crossing, 6 miles west of Pinedale, Wyoming, offered a stark example of how human-built infrastructure can interfere with migration routes.

Trappers Point is a narrow strip of land—shaped like the neck of an hourglass, with rivers on either side. The migrating pronghorn, "like sand falling through the hour glass," all crossed the highway in a single quarter-mile-wide area in the narrowest spot. Meanwhile, drivers reached the crest of a blind hill at 65 miles per hour just where the migration corridor intersected the highway.

To help mitigate the danger to drivers, the highway department put up signs that were supposed to flash when animals ran onto the highway. Even with the extra warnings, vehicles still collided with wildlife. After collar studies illuminated the migration's main threats, including the highway crossing at Trappers Point, conservation organizations and land-management agencies started to take actions that would span many years of work. In 2011, the Wyoming Department of Transportation designated $9.7 million to build wildlife-crossing structures at Trappers Point and a stretch of highway to the west. Engineer John Eddins proposed an overpass for antelope, which would justify the construction cost in a decade based on savings from averted car wrecks and wildlife deaths. In the first three years following construction, there were nearly 60,000 mule deer and 26,000 pronghorn crossings via this overpass. Traffic collisions with wildlife dropped more than 80 percent, from more than 135 in 2012 to fewer than 25 in 2015.

At 150 miles, the mule deer's Red Desert-to-Hoback migration is the longest, and one of the most difficult, of all the migrations in the Greater Yellowstone region; they cross five highways and must negotiate one hundred fences, as well as scale 11,000-foot mountains. All of their epic migration occurs outside national park boundaries.

Wildlife can thrive alongside human communities, as long as their migrations are given the chance to continue unimpeded.

Pronghorn won't jump fences, but they can go under those that have a bottom wire at least 16 inches off the ground.

Opposite: Fences are one of the key barriers to migrating pronghorn as they near Cody, Pinedale, and Jackson, Wyoming.
Above: Fences with barbless bottom wires that have sufficient space beneath them allow pronghorn to move through and not become tangled. With a little ingenuity, ranching and pronghorn can both thrive.

Arthur Middleton, Wes Livingston, and Lee Livingston studying maps in basecamp.

A solitary pronghorn fawn in Antelope Flats,
Grand Teton National Park

EPILOGUE

THOMAS LOVEJOY

AS A CHILD, I WOULD stare with fascination at a National Geographic Society globe, lit from within by an old-fashioned incandescent bulb. In particular, I was intrigued by the bright green places that were national parks—like the Everglades and Yellowstone—which then came to life in stories in *National Geographic Magazine.*

I understood these as places where nature reigned supreme and where the extraordinary variety of plants and animals (which today we call biodiversity) could be enjoyed and, to my naïve mind, would be "safe." That young impression certainly went beyond the impulses that created the earliest national parks. Those were mostly about preserving places of great beauty. In 1864, as the Civil War raged on, President Abraham Lincoln purchased and conveyed to the State of California the land that would later become Yosemite National Park. Then Old Faithful and the geological wonders of Yellowstone led to the creation (by President Ulysses S. Grant in 1872) of the world's first national park—Yellowstone National Park.

When these first national parks were created, Americans were occupied with notions of Manifest Destiny, busy bending nature to our purposes with little sense of the essential underpinning that nature provides to humanity. There was little recognition that Earth is a living planet where the biological and physical systems work together to cycle carbon, oxygen, and nitrogen and, in the process, determine the composition of the atmosphere and the climate. Surely, to the extent that conservation was important, our National Park System as initially conceived took care of that.

I was bound for a life of scientific adventure and was deep into that in the Brazilian Amazon when in 1968, a fellow graduate student showed me the newly published *Theory of Island Biogeography* by Robert MacArthur and Edward O. Wilson. The theory provided an elegant way to explain the different numbers of species on islands. Little did I realize that it would soon change my life and the world of conservation.

By the time I completed my PhD in 1971, a few farseeing scientists were publishing papers about how the idea of island biogeography might apply to remnant patches of nature and to designing and managing protected areas (not just in the United States but anywhere). Might such areas of remaining nature behave much like islands surrounded by water?

In their theory, MacArthur and Wilson focused on the implications of the different number of species between "continental islands" (which had been part of an adjacent continent when sea level was lower during the last ice age—like Trinidad), an equivalent area on the mainland, and "oceanic islands" (which had always been islands).

Generally speaking (with all aspects equivalent), an area of mainland has more species than a continental island of the same size. And that in turn has more species than a similar oceanic island. The implication was that once isolated by sea-level rise, the new islands lost species relative to the equivalent mainland area and trended toward the lower species number of oceanic islands. In other words, islands and thus habitats tend to lose species once they are not part of a larger expanse of habitat.

All of a sudden in the mid- to late 1970s, remnant patches of habitat were understood to be dynamic and losing species because they were no longer part of a larger habitat. The patch might be "safe," but all the constituent species were not. Habitat fragmentation became recognized as a serious conservation issue and an important part of the new science of conservation biology.

None of this was completely new, of course, but it did lead me in 1979 to initiate (with many colleagues) a giant experiment on forest fragmentation in the middle of the Brazilian Amazon to understand the process as well as the implications for conservation. Just four years later, in 1983, the Greater Yellowstone Coalition was founded in recognition that Yellowstone could not be successfully managed as an isolated entity. Rather, the national park needed to be managed in an integrated fashion with what we now call the Greater Yellowstone Ecosystem. The coalition invited me to address its annual meeting in 1987, to speak about the forest fragments' project.

About a decade later, in 2008, the forest fragment experiment was featured in a presentation to the Western Governors' Association, which focused on migratory animals and launched the Freedom to Roam initiative. A visionary program, it helps remove or modify barriers to migratory species, such as elk, mule deer, and pronghorn antelope. A deep appreciation for issues of habitat fragmentation and the need to restore natural connections in landscapes was in my mind when I first heard wildlife ecologist Arthur Middleton talk about his work on Yellowstone's elk and other migrations and when I saw some of Joe Riis's extraordinary photography.

In a related thread, the National Park Service—which celebrated its centennial in 2016 and is looking ahead into its second century—revisited a legendary report from the early 1960s by conservationist Aldo Leopold's son Starker that laid out the principles for management of national parks. No longer could parks be considered what the original report called "vignettes of primitive America," safe and secure like the green patches on my childhood globe. Rather, the Park Service's *Revisiting Leopold* highlights the need to view parks and protected areas as conservation anchors in larger conservation landscapes. It reaffirms the growing realization in science and conservation that nature needs connection as well as protection.

Lesson learned. But the idea of connection is pretty abstract. It is really about individual animals moving together in annual migrations essential for their survival, whether the migratory pathways taken by nine different elk herds to and from Yellowstone National Park, those of mule deer, or the movement of pronghorns—another fascination dating to my childhood. The extraordinary images of Joe Riis bring all this to life in ways that are beyond description: animals on the move, seeking, striving, experimenting, and surviving.

That all this science and art come together in this volume on the heels of the centennial year of the National Park Service is extraordinarily important. This book and similar efforts can help us reimagine and rethink conservation, help us move to a time when we truly recognize—in tangible ways—that human aspiration is embedded in nature. We can embrace nature, or we can ignore it and degrade it to our detriment.

The essays and photographs in this volume tell us clearly and strikingly that Yellowstone—the park but also the greater ecosystem in which it lies—is a model for how all of us should be thinking about and embracing nature everywhere. Nature is not secure within park boundaries, nor will humanity present and future be adequately served by persevering in such an illusion. The Yellowstone this volume reveals lights the way forward.

THE PHOTOGRAPHER'S STORY: QUIET PERSISTENCE

Photographer Joe Riis, something of an elusive wild creature himself, checking his wildlife camera trap

Up to four times the size of a mule deer,
elk weigh 400 to 1,000 pounds.

Opposite: Crossing a backwater channel of the Yellowstone River with pack mules loaded with camera gear and food
Above: Hall Sawyer, Arthur Middleton, and Joe Riis: their combination of rigorous academic study, adventurous storytelling, and stunning photography sheds new light on Yellowstone migrations.

Warmth and comfort for the researchers at the end of a long day

Above and opposite: Uncovering and documenting these migration routes was a journey and an adventure for the scientists and the photographer alike. Following animals that paid little heed to hiking trails posed route-planning challenges for Joe and Arthur.

THE FABLED THOROFARE PLATEAU

The Thorofare Plateau lies just outside the southeast corner of Yellowstone National Park. Commonly known as "The Thorofare," it is so-named because it serves as the main route for thousands of ungulates and other large animals who traverse it during their annual migrations.

The headwaters of both the Snake and Yellowstone rivers are found here. It is certainly the most remote swath of the Greater Yellowstone region, and arguably the most remote region in the continental United States.

BIBLIOGRAPHY

Berger, Joel, Steve Cain, and Kim Murray Berger. "Connecting the Dots: An Invariant Migration Corridor Links the Holocene to the Present." *Biology Letters* 2, no. 4 (2006): 528–31. doi:10.1098/rsbl.2006.0508.

Craighead, Frank C., Jr. "Grizzly Bear Ranges and Movement as Determined by Radiotracking." *Bears: Their Biology and Management* 3 (1976): 97–109. www.jstor.org/stable/3872759.

Craighead, John J., G. Atwell, and B. W. O'Gara. "Elk Migrations in and near Yellowstone National Park." *Wildlife Monographs* 29 (1972): 3–48.

Kauffman, Matthew J., James E. Meacham, Hall Sawyer, Alethea Y. Steingisser, William J. Rudd, and Emilene Ostlind. *Atlas of Wildlife Migration: Wyoming's Ungulates*, (in production). Corvallis, OR: Oregon State University Press, 2018.

Leopold, A. S., S. A. Cain, C. M. Cottam, I. N. Gabrielson, and T. L. Kimball. *Wildlife Management in the National Parks: The Leopold Report*. Report by the Advisory Board on Wildlife Management Appointed by Secretary of the Interior Udall, March 4, 1963. www.nps.gov/parkhistory/online_books/leopold/leopold.htm.

MacArthur, Robert H., and Edward O. Wilson. *The Theory of Island Biogeography*. 2nd ed. Princeton: Princeton University Press, 2001. First published 1967.

Marston, Richard, and Jay Anderson. "Watersheds and Vegetation of the Greater Yellowstone Ecosystem." *Conservation Biology* 5, no. 3 (1991): 338–46. doi:10.1111/j.1523-1739.1991.tb00147.x.

Merkle, J. A., K. L. Monteith, E. O. Aikens, M. M. Hayes, K. R. Hersey, A. D. Middleton, B. A. Oates, H. Sawyer, B. M. Scurlock, and M. J. Kauffman. 2016. "Large Herbivores Surf Waves of Green-up during Spring." *Proceedings of the Royal Society of London B* 283, no. 1833. http://dx.doi.org/10.1098/rspb.2016.0456.

Middleton, Arthur D., Matthew J. Kauffman, Douglas E. McWhirter, John G. Cook, Rachel C. Cook, Abigail A. Nelson, Michael D. Jimenez, and Robert W. Klaver. "Animal Migration amid Shifting Patterns of Predation and Phenology: Lessons from a Yellowstone Elk Herd." *Ecology* 94, no. 6 (2013): 1245–56. www.jstor.org/stable/23436142.

National Park System Advisory Board. *Revisiting Leopold: Resource Stewardship in the National Parks*. Washington, DC: National Park Service, 2012. www.nps.gov/calltoaction/PDF/LeopoldReport_2012.pdf.

Riis, Joe, and Jenny Nichols. *Elk River*. Short film. 2016. www.greateryellowstonemigrations.com/the-film.

Riis, Joe, and Tom Swartwout. *Mule Deer Migration*. Short video. 2014. https://vimeo.com/88619272.

Sawyer, Hall. *Seasonal Distribution Patterns and Migration Routes of Mule Deer in the Red Desert and Jack Morrow Hills Planning Area*. Laramie, WY: Western EcoSystems Technology, 2014. http://migrationinitiative.org/content/red-desert-hoback-migration-assessment.

Sawyer, Hall, Matthew Hayes, Bill Rudd, and Matthew Kauffman. *The Red Desert to Hoback Mule Deer Migration Assessment*. Laramie, WY: Wyoming Migration Initiative, University of Wyoming, 2014. http://migrationinitiative.org/content/red-desert-hoback-migration-assessment.

Sawyer, Hall, Matthew J. Kauffman, Arthur D. Middleton, Thomas A. Morrison, Ryan M. Nielson, and Teal B. Wyckoff. "A Framework for Understanding Barrier Effects on Migratory Ungulates." *Journal of Applied Ecology* 50 (2013): 68–78. doi:10.1111/1365-2664.12013.

Sawyer, Hall and Fred Lindzey. *Jackson Hole Pronghorn Study Final Report*. Laramie: Wyoming Cooperative Fish and Wildlife Research Unit, University of Wyoming, 2000. www.west-inc.com/reports/jackson_prongstudy.pdf.

Sawyer, Hall, Fred Lindzey, and Doug McWhirter. "Mule Deer and Pronghorn Migration in Western Wyoming." *Wildlife Society Bulletin* 33, no. 4 (2005): 1266–73. doi:10.2193/0091-7648(2005)33[1266:MDAPMI]2.0.CO;2.

US Congress. "An Act to Set Apart a Certain Tract of Land Lying Near the Head-Waters of the Yellowstone River as a Public Park." *United States Statutes at Large*, 42nd Cong., Sess. 2, Chap. 24, March 1, 1872. https://memory.loc.gov/cgi-bin/ampage?collId=llsl&fileName=017/llsl017.db&recNum=73.

White, P. J., Robert A. Garrott, and Glenn E. Plumb. *Yellowstone's Wildlife in Transition*. Cambridge, MA: Harvard University Press, 2013.

Arrowheads, chips, and flakes provide archaeological evidence of a long interconnection between humans and ungulates.

CONTRIBUTORS

Gretel Ehrlich is the author of thirteen books including *The Solace of Open Spaces*, *This Cold Heaven*, and *Facing the Wave*, nominated for a National Book Award. Her many honors include a Guggenheim, a PEN USA Award for Nonfiction, and the Henry David Thoreau Award for Nature Writing. She has lived on the eastern, southeastern, and northern edges of the greater Yellowstone ecosystem for forty-three years, and with Joe Riis, has followed the spring migrations of pronghorn, moose, elk, and mule deer.

Photo by Joe Riis

Photo by Slodoban Randjelovic

Thomas Lovejoy, an innovative conservation biologist, coined the term "biological diversity" (1980). In 2010 he was elected University Professor in the Department of Environmental Science and Policy at George Mason University. He is Senior Fellow at the United Nations Foundation based in Washington, DC.

Spanning the political spectrum, Lovejoy has served on science and environmental councils under the Reagan, Bush, and Clinton administrations. At the core of these many influential positions are seminal ideas, which have formed and strengthened the field of conservation biology. In the 1980s, he brought international attention to the world's tropical rainforests, and in particular, the Brazilian Amazon, where he has worked since 1965. With two co-edited books (1992 and 2005), he is credited as a founder of the field of climate change biology. He also founded the series *Nature*, the popular long-term series on public television. In 2001, Lovejoy was awarded the prestigious Tyler Prize for Environmental Achievement. In 2009 he was the winner of BBVA Foundation Frontiers of Knowledge Award in the Ecology and Conservation Biology Category. In 2009 he was appointed Conservation Fellow by National Geographic. In 2012 he was recognized by the Blue Planet Prize. Lovejoy holds BS and PhD (biology) degrees from Yale University.

Arthur Middleton is a wildlife ecologist who works on predator-prey interactions and migrations in large mammals. He is currently an assistant professor in the Department of Environmental Science, Management, and Policy at the University of California, Berkeley. His major, long-term field projects are located in the northern Rockies and the southern Andes. Much of his work has been focused in and around parks and protected areas, where he has sought to improve environmental outcomes by highlighting key differences between how systems work and how they are managed. Middleton has often stepped out of the normal science process to make creative products like museum exhibits and films, collaborating with photographers, filmmakers, and artists. Along with Joe Riis, he was awarded the 2013 Camp Monaco Prize by Prince Albert II of Monaco for linking research and public outreach on the subject of trans-boundary wildlife migrations in the Greater Yellowstone Ecosystem. He has a bachelor's degree from Bowdoin College, a master's degree from Yale University, and a PhD from the University of Wyoming. Middleton is originally from Charleston, South Carolina.

Photo by Joe Riis

Photo by Joe Riis

Emilene Ostlind is a Wyoming-based environmental journalist. She has contributed stories about wildlife, energy development, western communities, and natural resource issues to publications including High Country News, WyoFile.com, Wyoming Wildlife, the Patagonia blog The Cleanest Line, and others. From 2007 to 2009 she collaborated with Joe Riis to document and share the story of the western Wyoming pronghorn migration. Their High Country News cover story on the subject, "Perilous Passages," won the 2012 Science in Society Award from the National Association of Science Writers and the 2012 Knight-Risser Prize for Environmental Journalism in the West from Stanford University. Emilene is founding editor of Western Confluence magazine, a University of Wyoming publication covering natural resource science and management, and she is the text editor for the Wyoming Migration Initiative. She previously worked at National Geographic Magazine and High Country News. She grew up in Wyoming and holds an MFA in creative nonfiction writing with environment and natural resources from the University of Wyoming.

James Prosek is a distinguished artist, writer, and naturalist. His artwork has been exhibited at the Smithsonian American Art Museum, Tanya Bonakdar Gallery, Gerald Peters Gallery, the Dumbo Arts Center, and The Virginia Museum of Fine Arts, with solo exhibitions at The Aldrich Contemporary Art Museum, The Addison Gallery of American Art, the Philadelphia Museum of Art, and the National Academy of Sciences among others.

Prosek made his authorial debut at age nineteen, with *Trout: an Illustrated History*. He has also authored *Eels: An Exploration, from New Zealand to the Sargasso, of the World's Most Amazing and Mysterious Fish*, which was featured in the PBS series *Nature*, and *Ocean Fishes*, a collection of life size paintings of thirty-five Atlantic fishes. He won a Peabody Award for his documentary about traveling in the footsteps of Izaak Walton. The Academy of Natural Sciences in Philadelphia awarded him with the Gold Medal for Distinction in Natural History Art in 2012. With Patagonia owner Yvon Chouinard, Prosek co-founded World Trout, a conservation initiative which supports cold water habitat conservation. Prosek is a curatorial affiliate at Yale's Peabody Museum of Natural History and serves on the Yale Institute for Biospheric Studies advisory board.

Photo by Joe Riis

PHOTOGRAPHER

Joe Riis is a National Geographic Fellow and Wyoming Migration Initiative (WMI) Photography Fellow. Trained as a wildlife biologist, Joe is known for his pioneering, award-winning photography of animal migrations in the West, often using camera traps to capture up-close portraits of migrating wildlife. He has worked on wildlife assignments for *National Geographic Magazine* on five continents. His photographs have been exhibited widely and are included in private and public collections worldwide.

In 2008–09, Joe followed the path of the pronghorn from the Green River Basin to Grand Teton National Park, documenting in photographs and video this migration phenomenon for the first time. Joe later campaigned for eight wildlife overpasses and underpasses now built by the Wyoming DOT to help maintain the migration corridor. Along with writer Emilene Ostlind, he was awarded the Stanford Knight-Risser Award for western environmental journalism. He earned an Emmy award for his pronghorn cinematography on the National Geographic *Great Migrations* series.

Joe often partners directly with research scientists to tell the science story to the greater public through all media. In 2012–13, Joe worked with researcher Hall Sawyer and the WMI to document the Red Desert to Hoback mule deer migration in Wyoming and produced a short film of the migration that has garnered nearly 5 million views on YouTube. In 2014–15, Joe worked with researcher Arthur Middleton to document the elk migrations of the Greater Yellowstone, work that was awarded the inaugural Camp Monaco Prize from Prince Albert II of Monaco and the Buffalo Bill Center of the West. Additionally, Arthur and Joe were named National Geographic Adventurers of the Year in 2016.

Joe lives in rural South Dakota near the Missouri River.

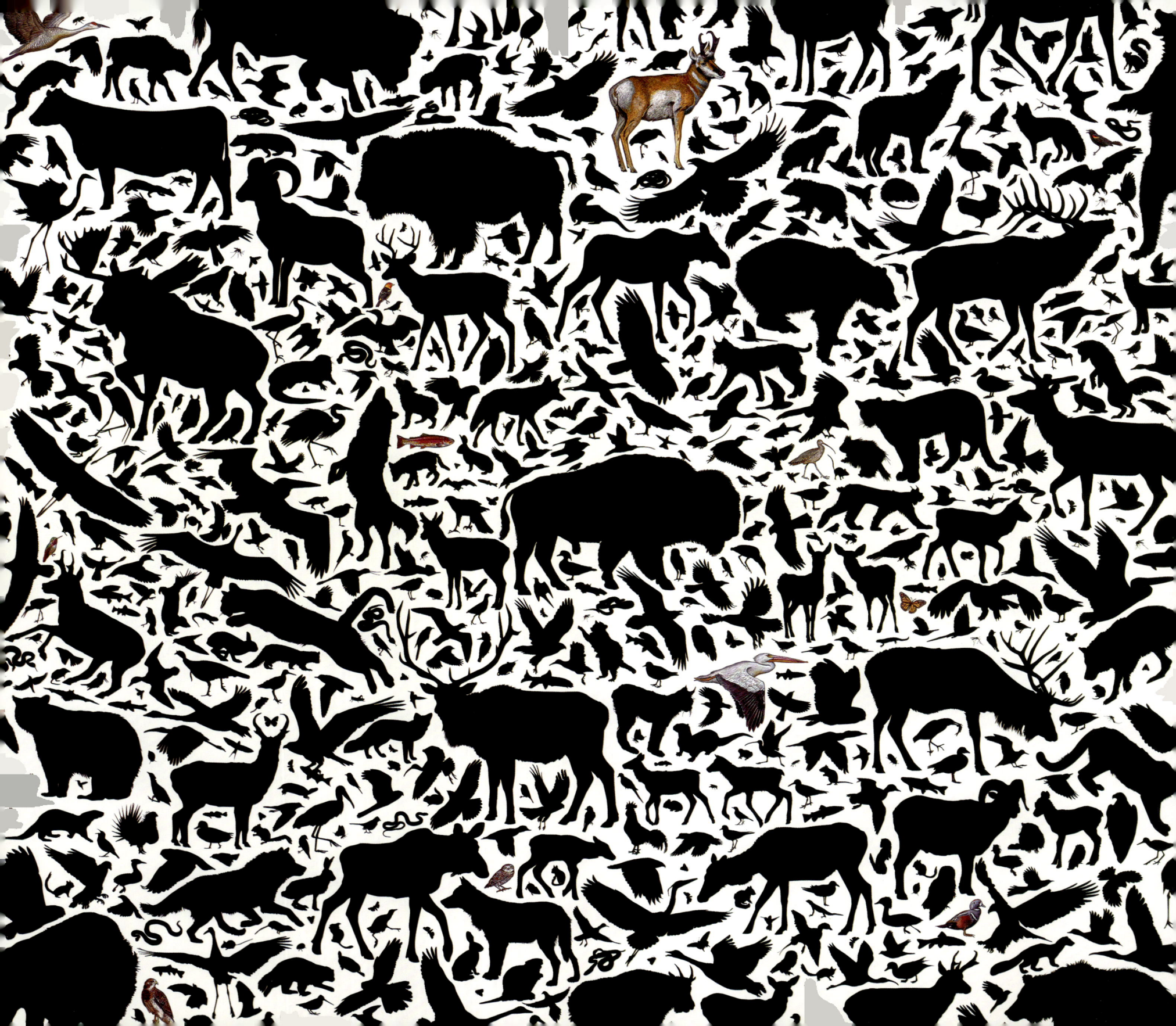

IN APPRECIATION

Support for the creation of this book came from the George B. Storer Foundation, the Weeden Foundation, the Bobby Model Charitable Fund, and the Cinnabar Foundation.

Additional support was provided by the individuals in Braided River's Headwaters Club—Tom and Sonya Campion, Ann and Ron Holz, Jacob and Margo Engelstein, Craig McKibben, and Sarah Merner.

Braided River's primary nonprofit partner for this book is the Wyoming Migration Initiative, whose mission is to advance the understanding, appreciation, and conservation of Wyoming's migratory ungulates by conducting innovative research and sharing scientific information through public outreach.

These organizations provided Joe Riis principal support for his years of pronghorn fieldwork: National Geographic Expeditions Council, University of Wyoming Larsh Bristol Photojournalism Fellowship, and the North American Nature Photography Association's Philip Hyde Environmental Grant. And for his mule deer fieldwork: the George B. Storer Foundation and the Knobloch Family Foundation.

The following provided Joe Riis and Arthur Middleton with invaluable support throughout their years of elk photographic fieldwork and scientific research: the Prince Albert II of Monaco Foundation, the Draper Museum of Natural History at the Buffalo Bill Center of the West, the Biodiversity Institute at the University of Wyoming, the George B. Storer Foundation, the Knobloch Family Foundation, the Fran and Lenox Baker Foundation, the National Geographic Expeditions Council, Shoshone National Forest, the Duncan Fund, and Mary Anne and Bill Dingus. Throughout this effort, their primary partners were the Wyoming Migration Initiative, the Yale School of Forestry and Environmental Studies, and the Buffalo Bill Center of the West.

Federal, state, university, and NGO partners who shared data used in this book include the Wyoming Game and Fish Department; Montana Fish, Wildlife, and Parks; Idaho Department of Fish and Game; US National Park Service; US Fish and Wildlife Service; Iowa State University; and the Wildlife Conservation Society.

James Prosek
Yellowstone Composition No. 1
Whitney Western Art Museum,
Buffalo Bill Center of the West

ACKNOWLEDGMENTS

It took me nearly a decade to make all the photographs that are in this book, and the project continues. My long-term approach means moving to the community where I'm photographing and filming. I like to get to know not only the places but also the people who call them home and that takes continued presence and letting the seasons, movements, and interactions unfold in their own time. It was my dream to watch and document the Teton pronghorn migration and get to know the people who live in and around the corridor. That migration led me to the mule deer migration, which then led me to the elk migration. It was all part of my Yellowstone Migrations project, a project that will likely continue my entire life. I have many people to thank for introducing me to and supporting me in what has become my calling.

Thank you Emilene Ostlind, Hall Sawyer, Doug McWhirter, Karsten Heuer, Leanne Allison, Florian Schulz, Michael Soulé, and Rick Ridgeway for opening my mind to the world of animal migrations. You have inspired me to do this work.

Thank you to the Domek Family for always having a warm place for me to stay and eat when I "stop by"; you have become like a second family to me. Thanks to my longtime friend Jeff Jewell for help in the field and to Steve Duerr and Kathleen Belk from The Murie Center for providing the cabin that I used as my basecamp in 2009. For my aerial photography, thanks to Chris Boyer and LightHawk. To Gretel Ehrlich for allowing me to live in her cabin in 2012–13, and for the adventures since then—thank you and I'm looking forward to more! Thanks also to Barron Collier for providing a place to stay and organize gear for three years during the elk project. A special thank you to Matt Kauffman, Jeff Parrish, Matt Wagner, Jane Sievert, Anne Young, Faith Model, Steve Winter, Doug Chadwick, Paul Klingenstein, Kathy Bole, Lee Livingston, Wesley Livingston, Wes Livingston, and Scott LaFevers. And thanks especially to Rebecca Martin, Kathy Moran, Michael "Nick" Nichols, and Chris Johns at the National Geographic Society and to Liz Storer for believing in this work from the start.

Arthur Middleton, thank you for the friendship and collaboration that made our elk project a success and a lot of fun. I look forward to more days on the trail with you.

And to the writers of this book, some of my closest friends: Emilene, Gretel, Arthur, and Tom Lovejoy, thank you for your prose—you bring these migrations to life. And at Braided River, thank you to Helen Cherullo, Deb Easter, Lace Thornberg, and Mary Metz for making this book possible. I appreciate your commitment and unwavering focus; you have been wonderful partners in this book.

There have been many people who have helped me over the years; you know who you are. From helping me with flat tires on the side of the highway to sharing what you've seen in the backcountry or frontcountry, for inspiring the photographs in this book, thank you.

To my parents Jim and Jeanne, who introduced me to the wild planet at a young age and have always supported me with love, time, and kindness, you have been the best parents a kid could ever ask for, with love, thank you. And lastly, to my partner Lucy, I love you.

Elk on winter range in the Pitchfork Ranch in Wyoming

JAMES PROSEK'S "YELLOWSTONE COMPOSITION NO. 1"

"The idea of an ecosystem itself is flawed. The existence of the word suggests that an ecosystem has a beginning and an end; the word puts up walls.... We have yet to truly and finally embrace… a holistic, interconnected planet."– James Prosek

Yellowstone Composition No. 1 (Whitney Western Art Museum, Buffalo Bill Center of the West) highlights twelve of Yellowstone's long-distance migrants. Prosek depicted these twelve creatures in color amongst almost seven hundred others shown in silhouette to suggest interconnectedness between these migrant species and Yellowstone's residents. The work also suggests that Yellowstone's wildlife require freedom to roam beyond the park's boundaries. Image courtesy of James Prosek and Schwartz-Wajahat, New York.

MAPS

The maps in this book have been generously donated by the Wyoming Migration Initiative (migrationinitiative.org) and created by cartographers at the University of Oregon. These maps, and many others illustrating the wildlife migrations in Wyoming, will be included in *Atlas of Wildlife Migration: Wyoming's Ungulates*, by Matthew J., Kauffman, James E. Meacham, Hall Sawyer, Alethea Y. Steingisser, William J. Rudd, and Emilene Ostlind to be published in 2018 by Oregon State University.

CAPTIONS FOR OPENING CLOSING PORTFOLIOS

page 1 A band of pronghorn does migrate through the upper Green River Basin; *title page* Cow elk swimming the South Fork of the Shoshone River pulsing from spring snowmelt; *page 4* Joe packing a mule with his photo gear on the Thorofare Plateau (photo by Shane Moore); *pages 6–7* During their fall migration, as they move from the mountains into the desert, mule deer will cover hundreds of miles; *pages 8–9* Four wolves walking the ridge line near an elk wintering range in the Shoshone National Forest; *pages 10–11* Pronghorn making their way through the Gros Ventre River drainage; *pages 12–13* Deep winter snows place a burden on pronghorn; *pages 14–15* Mule deer approach a ridgeline during their fall migration; *pages 16–17* Three-week-old elk calves making their first migration to their summer range in southeastern Yellowstone; *pages 18–19* A migrating doe pronghorn tried to jump over a fence and caught her leg instead. Joe was on hand to pull this fence apart to free her. The section of fencing has since been replaced with a wildlife-friendly fence; *Table of Contents* A group of bull elk in June with antlers still in their velvet stage; *page 176* Cow elk and calf

BRAIDED RIVER

BRAIDED RIVER, the conservation imprint of Mountaineers Books, combines photography and writing to bring a fresh perspective to key environmental issues facing western North America's wildest places. Our books reach beyond the printed page as we take these distinctive voices and vision to a wider audience through lectures, exhibits, and multimedia events. Our goal is to build public support for wilderness preservation campaigns, and inspire public action. This work is made possible through the book sales and contributions made to Braided River, a 501(c)(3) nonprofit organization. Please visit www.braidedriver.org for more information on events, exhibits, speakers, and how to contribute to this work.

Braided River books may be purchased for corporate, educational, or other promotional sales. For special discounts and information, contact our sales department at mbooks@mountaineersbooks.org or 800.553.4453.

THE MOUNTAINEERS, founded in 1906, is a nonprofit outdoor activity and conservation organization, whose mission is "to explore, study, preserve, and enjoy the natural beauty of the outdoors . . ." Mountaineers Books supports this mission by publishing travel and natural history guides, instructional texts, and works on conservation and history. Send or call for our catalog of more than 700 outdoor titles:
Mountaineers Books
1001 SW Klickitat Way, Suite 201
Seattle, WA 98134
800.553.4453
www.mountaineersbooks.org

Manufactured in China on FSC-certified materials, using soy-based ink.

© 2017 by Braided River
All rights reserved.
First edition, 2017

No part of this book may be reproduced in any form, or by any electronic, mechanical, or other means, without permission in writing from the publisher.

Publisher: Helen Cherullo
Project Manager: Mary Metz
Acquisitions and Developmental Editor: Deb Easter
Content and Copy Editor: Julie Van Pelt
Cover and Book Designer: Heidi Smets
Development and Communications: Lace Thornberg
Cartographers: Alethea Steingisser, James Meacham, and Dylan Molnar, University of Oregon Department of Geography
Scientific Advisor: Hall Sawyer
Illustration by James Prosek

Library of Congress Cataloging-in-Publication Data
Names: Riis, Joe.
Title: Yellowstone migrations / Joe Riis.
Description: First edition. | Seattle : Braided River, 2017. | Includes bibliographical references.
Identifiers: LCCN 2017009758 | ISBN 9781680510898
Subjects: LCSH: Animal migration—Yellowstone National Park—Pictorial works.
 | Mule deer—Migration—Yellowstone National Park—Pictorial works. | Pronghorn—Migration—Yellowstone National Park—Pictorial works. | Elk—Migration—Yellowstone National Park—Pictorial works. | Wildlife photography—Yellowstone National Park.
Classification: LCC QL754 .R55 2017 | DDC 591.56/80978752—dc23
LC record available at https://lccn.loc.gov/2017009758